SOCIAL EVOLUTION AND INCLUSIVE FITNESS THEORY

SOCIAL EVOLUTION AND INCLUSIVE FITNESS THEORY

An Introduction

James A. R. Marshall

Princeton University Press
Princeton and Oxford

Published by Princeton University Press, 41 William Street,
Princeton, New Jersey 08540
In the United Kingdom: Princeton University Press, 6 Oxford Street,
Woodstock, Oxfordshire OX20 1TR
press.princeton.edu

Cover photograph by Christopher Tranter.
Cover design by Wanda España.

First paperback printing, 2019

Paper ISBN 978-0-691-18333-6

The Library of Congress has cataloged the cloth edition as follows:

Marshall, James A. R., 1976–
Social evolution and inclusive fitness theory : an introduction / James A.R. Marshall.
    pages cm
Includes bibliographical references and index.
ISBN 978-0-691-16156-3 (alk. paper)
1. Sociobiology.   2. Social behavior in animals.   3. Behavior evolution.
4.   Evolution (Biology)   5. Social evolution.   6. Hamilton, W. D.
(William Donald), 1936-2000. I. Title.
GN365.9.M37 2015
304.5–dc23   2014031810

British Library Cataloging-in-Publication Data is available

This book has been composed in Garamond Premr Pro and Avenir LT Std
Typeset by S R Nova Pvt Ltd, Bangalore, India

For Ruth, Hannah, and Adam

CONTENTS

LIST OF ILLUSTRATIONS

FIGURES

TABLES

PREFACE

Inclusive fitness theory is probably the most important advance in our understanding of evolution since 1859. Darwin and Wallace's[1] initial insight, that natural selection could explain adaptation and variation, rested on the assumption that natural selection acts on direct individual reproduction. This assumption remained intact during the mathematical formalization of natural selection theory during the modern synthesis of the 1930s, particularly in the seminal work of R. A. Fisher. Yet if natural selection acts on direct individual reproduction, a fundamental problem arises for the theory: how to explain apparent incidents of self-sacrifice in reproduction by individuals, for the benefit of the reproduction of others? For a while, some researchers assumed the problem could be solved by considering groups as the unit of selection, in which case individuals might sacrifice their own reproduction to benefit that of the group as a whole. However, this "naïve group selection" was shown to be logically inconsistent. The true resolution of the apparent paradox came in 1963 and 1964 with the publication by William D. Hamilton of the theory of inclusive fitness, according to which individuals should value not only their own reproduction, but also that of genetically related individuals. Hamilton thus extended classical Darwin–Wallace–Fisher fitness to a more inclusive version, able to explain the evolution of self-sacrifice.

The goal of this book is to celebrate the 50th anniversary of the publication of Hamilton's inclusive fitness theory by providing a relatively brief introduction to, and explanation of, its generality and its correctness. This is necessary, because both the generality and correctness of inclusive fitness theory have been repeatedly

---

[1] On the centenary of his death it is particularly fitting to note Wallace's often neglected role in conceiving evolution through natural selection.

criticized since its inception, and the pace and profile of these criticisms has increased in recent years. The situation for a researcher making their first forays into social evolution theory is daunting. The literature is vast and, given the subtleties of the theory and its apparently controversial status, it can seem difficult to know where to start and what to believe. Large literatures usually require effective summarization in books, and several excellent books in social evolution already exist. Yet all these books achieve something slightly different to the aims of this one. Some are conceptually and technically deep, but may be correspondingly difficult to approach. Others provide a more accessible introduction to the relevant theory, but at the expense of some of the technical depth. Still others survey empirical data on social evolution, and interpret it through the lens of theory. This book aims to fall somewhere between these three extremes, providing a brief primer in the basic concepts of social evolution theory, including its major subtleties and its empirical application, while being simultaneously accessible and technically complete.

What will a reader find in this book, and what will they not? In this book I focus on trying to expose the logical core of inclusive fitness theory, in a way that is understandable to the less mathematically confident, while still providing enough of the technical detail for those who are interested to follow up on. By trying to get to the level of causal explanations of the evolution of social traits, I hope to be able to explain why controversies and misunderstandings have arisen. Explaining how divergent views arise is often a good route towards their speedy resolution. Dealing with causation is, of course, a difficult philosophical problem. The tool I have chosen primarily to base the explanations in this book on is quantitative genetics. Quantitative genetics has been used extensively in studying evolution in general, and many believe it provides a useful route to understanding social evolution. Quantitative genetics is an inherently statistical framework, able to deal with complicated phenotypes, with poorly understood underlying genetics, and still to draw meaningful conclusions. Hence there are no complicated population genetical models in this book. Similarly, popular approaches to developing maximization arguments using concepts of evolutionary stability, which require weak selection, do not feature here. Modeling approaches necessarily trade predictive power against generality, and the focus of this book is on showing the generality of inclusive fitness theory. Because of the minimal assumptions it makes, quantitative genetics has very broad applicability, and the conclusions it draws should hold in great generality. For identifying and explaining the general principles of social evolution theory, it thus seems ideal.

This book was written during the 50th anniversary year of Hamilton's first outline of inclusive fitness theory, and I write these words on the eve of the 50th anniversary year of the first full mathematical presentation of that theory. In the last five decades inclusive fitness has grown from initial obscurity to being generally recognized as the fundamental quantity that natural selection acts on. We should look forward to the even deeper understanding of biological systems that 50 more years of the application of inclusive fitness theory will certainly provide us with.

James A. R. Marshall
Sheffield, December 2013

ACKNOWLEDGMENTS

Many people have helped to make this book possible, to improve it substantially, or both. First I must thank Tim Clutton-Brock for advice and for putting me in touch with my publisher, Princeton University Press. At the Press I thank the staff, but particularly Alison Kalett for taking a chance on a new and untried author.

In writing this book, and over the years before, I have had useful discussions with a number of people. These include Marco Archetti, Jonathan Birch, Chris Cannings, Tim Clutton-Brock (with particular thanks for a stimulating visit to the Kalihari), Jeff Fletcher, Kevin Foster, Andy Gardner, Bill Hughes, Loeske Kruuk, John McNamara, Samir Okasha, Dave Queller, Corina Tarnita, and Geoff Wild.

Several colleagues were kind enough to provide very useful comments on draft chapters, or parts thereof. These are Marco Archetti, Jonathan Birch, Andrew Bourke, Tim Clutton-Brock, A.W.F. Edwards, Kevin Foster, Steve Frank, Andy Gardner, Melanie Ghoul, Ashleigh Griffin, John McNamara, Samir Okasha, Wenying Shou, and Stu West.

Special thanks are due to Ben Hatchwell, Chris Quickfall, Andy Gardner, and Stu West, who all read the entire manuscript draft and always provided insightful suggestions. Chris Quickfall, in particular, checked and rederived all of my mathematics, and Andy Gardner caught what were hopefully the final few mistakes. Any remaining errors are my own; I have made every effort to avoid these, but should any reader detect what they think is an error, I ask them to contact me.

I should also like to thank the photographers who kindly allowed me to use their images. Photographs are credited where they appear, but here I thank again Artour Anker, Kevin Foster, Melanie Ghoul, Loeske Kruuk, and Chris Tranter. I also thank Tim Clutton-Brock and Amanda Ridley for providing me with the opportunity to take some photographs for the book myself. Finally, Loeske Kruuk was kind enough to provide the source data enabling figure 3.1 to be drawn.

SOCIAL EVOLUTION AND INCLUSIVE FITNESS THEORY

CHAPTER ONE

Social Behavior and Evolutionary Thought

1.1 EXPLANATIONS FOR APPARENT DESIGN

Animals, plants, and other organisms appear to be designed for some purpose. While the ultimate purpose may not always be clear to us, observers of the natural world can readily understand sophisticated "devices" such as the wing and the eye to be "designed" for flight and sight, respectively. Until the mid-nineteenth century, natural philosophy explained design in nature as being due to, and evidence for, the existence of a supernatural creator. One of the most famous late examples of this tradition is William Paley's "argument from design" [Paley, 1802]; on discovering a pocket watch lying on a heath, the conclusion of any reasonable person is that, due to its apparent complexity and its evident purpose, it must have been designed, and therefore a designer (the watchmaker) must exist. Paley went on to argue that, should the discovered watch have an internal mechanism capable of producing copies of itself, the rational discoverer would still conclude that it had been designed for this purpose, in addition to its purpose of telling the time, and must still therefore have a designer. Similarly, the apparent complexity in construction of animals and plants, and fitness for a purpose which includes reproduction, means they must have been designed, and therefore a designer (God) must exist. Under such a view, of course, an anthropocentric natural theologist might conclude that the animals and plants around us have been designed, by the supernatural creator, with the primary purpose of giving us food to eat, natural resources with which to make things, and so on.

With the work of Charles Darwin and of Alfred Russel Wallace [Darwin and Wallace, 1858, Darwin, 1859], an alternative explanation for the appearance of design arrived and, simultaneously, the question of the ultimate purpose of organisms was answered. The ultimate purpose of organisms was to compete for individual reproduction, and the result of such competition was that natural selection would progressively improve their suitability for this purpose, thereby giving them the appearance of design. If flight would increase the chances of individual reproduction for members of a species, for example, then natural selection acting on heritable variation over many generations could fashion limbs into wings, and then progressively optimize them for the purposes of aerodynamically efficient flight. Design and purpose in nature were both explained, and the explanations did not suggest a supernatural designer.

Darwin and Wallace amassed significant empirical support for the theory of evolution through natural selection, in collections of animals from around the globe,[1] and Darwin also interacted with practitioners of artificial selection, such as pigeon breeders and farmers. Yet the new evolutionary theory was formulated without knowledge of how characteristics, which natural selection was supposed to act on, were inherited by offspring from their parents. In fact, only 8 years after Darwin and Wallace's papers were read at the Linnean Society in London, Gregor Mendel discovered the particulate nature of inheritance in an abbey in Brno, through his experiments on pea morphology [Mendl, 1866]. Despite being contemporary with and crucially relevant to the theory of natural selection, Mendel's results were ignored for over 30 years [Bateson, 1909]. Initially thought to be a replacement for Darwinian evolution, the field of genetics was ultimately reconciled with natural selection in a mathematical framework that came to be known as the "modern synthetic theory of evolution," or "modern synthesis" for short [Huxley, 1942]. Primarily the work of three pioneers, Sewall Wright, J.B.S. Haldane, and R. A. Fisher (e.g., [Wright, 1932, Haldane, 1932, Fisher, 1930]), the modern synthesis gave a formal mathematical structure to Darwin and Wallace's ideas that would enable them to be developed into a predictive theory as never before. Of particular importance, in *The Genetical Theory of Natural Selection* Fisher mathematically formalized individual reproductive success, which lies at the original heart of natural selection theory [Fisher, 1930]. Thus, with a few exceptions as discussed below, in explaining adaptation the modern synthesis firmly set the focus of natural selection at the level of the individual and their own direct reproduction.

1.2 NATURAL SELECTION AND SOCIAL BEHAVIOR

Although the examples described above of traits "designed" through natural selection are physical body parts, behaviors also have genetic components, and therefore can be shaped by natural selection. As William D. Hamilton put it very pithily, "It is generally accepted that the behaviour characteristic of a species is just as much the product of evolution as the morphology" [Hamilton, 1963]. Behaviors that improve the reproductive success expected by an individual often have a negative impact on reproduction of members of the same species; one obvious example is behaviors involved in competition over mates, such as in display and fighting by red deer stags (figure 1.1); by monopolizing access to females, a male improves his own reproductive success at the expense of other males. Natural selection theory as developed by Darwin, Fisher, and others has no problem explaining the evolution of such behaviors; indeed it predicts them. This theory acts according to the reproductive success of individuals, and when the side effects of any trait are to modify the reproductive success of unrelated individuals, these are irrelevant.

Other individual behaviors seem to impact on the reproduction of others in a much more "deliberate" manner, however. Examples of such *social behaviors* abound in the natural world. Quite possibly the most well-known examples are among the social insects, considered by Darwin himself [Darwin, 1859]. In these insect species, reproductive division of labor is observed, with one or more castes helping to raise offspring other than their own; this is referred to as *eusociality* [Crespi and Yanega, 1995]. The simplest pattern is that the daughters of a single reproductive female, the queen, forage for, defend, and raise her offspring. These worker daughters either have suppressed levels of reproduction, as in the honeybee *Apis mellifera* where workers may both reduce their own levels of reproduction and destroy eggs laid by other workers [Ratnieks and Visscher, 1989], or are completely functionally sterile, as in several species of leafcutter ant for example (figure 1.2). *Cooperative breeding* is also observed in vertebrates, including many species of birds (e.g., figure 1.3) and mammals, such as meerkats (*Suricata suricatta*; e.g., [Clutton-Brock et al., 1998]) and naked mole rats (*Heterocephalus glaber*; [Jarvis, 1981]).[2] Cooperative breeders exhibit similar behaviors to eusocial species, in that helpers forage for, and guard, the offspring of a single breeding pair, although helpers do not form a distinct caste and may subsequently become reproductives themselves [Crespi and Yanega, 1995]. The presence of helpers has been shown to improve reproductive success by the breeding pair (e.g., [Hatchwell et al., 2004]),

Figure 1.1: A red deer stag (*Cervus elaphus*). Stags possess large antlers which impact on performance in fights, dominance rank, and hence access to fertile females. Maintenance of a harem of females, and hence increased reproductive success, negatively impacts on the reproductive success of other males in the population. However, natural selection theory (as developed by Darwin, Fisher, and others) acting on individuals explains the evolution of antlers since the successful male's net personal reproduction is increased as a result of having them. Photograph by Loeske Kruuk, reproduced from [Kruuk et al., 2014] with the permission of the photographer.

yet the helpers necessarily forego their own reproduction while caring for offspring that are not their own (e.g., [Emlen, 1982]).

Less frequently appreciated, social behavior is also observed in microbes including amoebae and bacteria [West et al., 2007a]. In social amoebae (*Dictyostelium* sp.; figure 1.4) normally free-living individuals aggregate at times of ecological stress, with some amoebae sacrificing themselves to form a structure that raises other individuals up in order to facilitate their dispersal to new, potentially richer,

Figure 1.2: Leafcutter ants of the genus *Atta* have morphologically distinct worker castes [Wilson, 1980], such as this forager (carrying leaf) and minim (sitting on leaf). In eusocial insect colonies, a worker caste or castes are either partially or totally functionally sterile, reducing or foregoing individual reproduction in order to support the reproduction of their mother. *Atta colombica* workers, although still possessing functioning ovaries, are effectively sterile [Dijkstra et al., 2005, Dijkstra and Boomsma, 2006]. Photograph by Chris Tranter, reproduced with permission of the photographer.

locations (figure 1.4) [Raper, 1984]. In the bacterium *Pseudomonas aeruginosa*, as in many other microorganisms, individuals secrete siderophores, which scavenge iron from insoluble forms in the environment. Siderophore production is individually costly in metabolic terms, resulting in a reduced growth rate, but this is offset by the increase in growth rate that siderophores facilitate when iron is scarce [Griffin et al., 2004, Jiricny et al., 2010] (figure 1.5A). However, siderophores secreted by individual bacteria can also facilitate iron uptake by neighboring individuals, allowing them to benefit from an increased growth rate, even if those neighbors did not contribute to siderophore production themselves [Griffin et al., 2004, Jiricny et al., 2010] (figure 1.5B).

Less munificent examples of social behavior have also been described. Let us take one important example: bacterial production of bacteriocins. Production of colicins by the bacterium *Escherichia coli*, for example, is fatal for producing cells, as well as killing neighboring cells within a narrow phylogenetic range [Riley and Wertz, 2002]. Thus, bacteriocin production is personally costly (colicin producers pay the ultimate price of death, thereby ceasing personal reproduction),

Figure 1.3: A long-tailed tit chick (*Aegithalos caudatus*) is fed an invertebrate by an adult helper. Long-tailed tits are facultatively cooperative breeders; individuals that have failed to breed successfully themselves may help to raise the offspring of another breeding pair by feeding their chicks until independence [Hatchwell and Sharp, 2006, Hatchwell et al., 2014]. Photograph by Ben Hatchwell, reproduced with permission of the photographer.

as well as costly to the targets of the behavior. Similarly, *Pseudomonas* bacteria produce individually costly pyocins that inhibit growth of strains that do not possess corresponding immunity genes [Michel-Briand and Baysse, 2002], as illustrated in figure 1.6.

Some purely physical traits can also have positive or negative social effects on conspecifics. One example is aposematism, for example in caterpillars (figure 1.7), in which individuals evolve both to be unpalatable to predators, and to bear conspicuous markings that indicate their unpalatability. As described below, Fisher himself considered the problem of aposematism, the initial evolution of which would be personally costly since conspicuous markings increase the probability of detection by a predator. A conspicuous distasteful individual being consumed would benefit aposematic members of the same species, however, by informing the predator that conspicuous markings mean unpalatability and thereby deterring them [Fisher, 1930].[3] Thus, while many social effects on conspecifics are due to behavior, not all are.

Examples such as these, and many others, have long presented a puzzle for evolutionary biology. The puzzle is that, under Darwin's and Fisher's views of

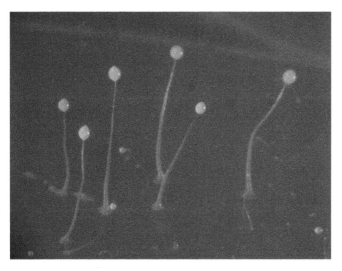

Figure 1.4: Various social amoebae such as *Dictyostelium discoideum* exhibit primitive multi-cellularity under certain conditions. When resources are locally depleted, free-living amoebae aggregate to form a multicellular "slug," which migrates and then produces a stalk topped by a fruiting body. Only amoebae in the fruiting body become spores and therefore potentially reproduce; all members of the stalk die in raising the fruiting body high enough to disperse spores effectively [Raper, 1984]. Photograph by Kevin Foster, reproduced with permission of the photographer.

natural selection acting on personal reproduction, it seems to make no evolutionary sense for individuals to reduce their personal reproductive success, possibly to zero, in order to have an effect on the reproduction of others. Natural selection should favor traits that *increase* personal reproductive success, hence personally costly traits of the kind described above should experience negative selection, and be eliminated from any population in which they appear. Yet the examples we have just seen seem not to be of transient social behaviors in the process of being weeded out by natural selection, but rather of stable evolutionary outcomes; eusociality has evolved multiple, independent times in the social insects, for example, and persisted for millions of years (e.g., [Hughes et al., 2008]). The crucial question, therefore, is how can evolutionary theory be extended to accommodate these obvious biological facts?

1.3 ARGUMENTS FOR GROUP BENEFIT

Darwin himself was concerned with explaining the evolution of apparent self-sacrifice, commenting on sterility in the social insects in *On the Origin of Species*

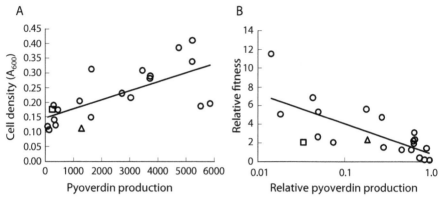

Figure 1.5: Fitness consequences of production of a siderophore (pyoverdin) in *Pseudomonas aeruginosa*. (A) Cell density in monocultures of different *P. aeruginosa* mutant strains in iron-limited environments, plotted against strain pyoverdin production per cell; when soluble iron is scarce, pyoverdin production positively correlates with higher cell densities, indicating that it is beneficial for population growth. (B) Relative competitive ability of different *P. aeruginosa* mutant strains in competition with the wild type, plotted against relative strain pyoverdin production; the negative correlation indicates that pyoverdin production is individually costly. In both figures, plot markers indicate mutant strains having different provenance. Figures redrawn from [Jiricny et al., 2010].

[Darwin, 1859]. Interestingly, Darwin was not concerned with sterility per se, which he considered could be explained by selection acting at the level of the family (i.e., colony), but rather how morphologically varied worker castes could evolve when they left no offspring to inherit their variation [Ratnieks et al., 2010]. Darwin resolved this problem by analogy to the selection of well-flavored vegetables or beef cattle. Once a vegetable or a cow has been consumed and found to be tasty it is unavailable for production of further similarly tasty individuals; yet horticulturists and beef farmers had long known that by breeding from closely related individuals in the vegetable patch or herd, offspring with similar characteristics could be produced [Darwin, 1859, Ratnieks et al., 2010].

Just over 10 years later, in *The Descent of Man*, Darwin discussed the potential for competition between human groups to favor those groups containing more "courageous, sympathetic and faithful members, who were always ready to warn each other of danger, to aid and defend each other." Darwin noted that in intergroup conflict, groups with the highest numbers of such individuals would tend to be victorious, all else being equal. Yet he also noted that within those same groups, the same such individuals would be at a disadvantage; "therefore it hardly seems

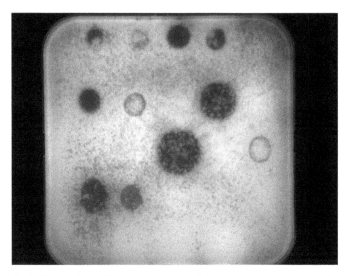

Figure 1.6: A "bacterial lawn" of a single strain of *Pseudomonas aeruginosa*, isolated from a pulmonary infection. Supernatant spots, containing pyocins, from 25 environmental and clinical strains of *P. aeruginosa* have been imposed on the lawn. Where growth of the lawn is inhibited under a spot (dark disks), this is due to action of the pyocins in the spot, and lack of appropriate pyocin-immunity genes in the lawn [Michel-Briand and Baysse, 2002]. Photograph by Melanie Ghoul, reproduced with permission of the photographer.

probable, that the number of men gifted with such virtues, or that the standard of their excellence, could be increased through natural selection, that is, by the survival of the fittest" [Darwin, 1871].

Darwin understood that, technically, selection could act at multiple levels in a biological hierarchy. Samir Okasha [Okasha, 2006] traces the full development of these ideas back to August Weismann [Weismann, 1903]. Darwin, however, also noted the fundamental difficulty with selection acting, in the way he described, at a higher level than the individual; while between-group selection would favor groups containing more individuals with prosocial behaviors, defined as behaviors that benefit others, within the group these behaviors would be individually costly and hence disfavored by natural selection at the level of the individual.

Despite Darwin's clear statement of the fact that individual selection would run counter to selection acting at higher levels, by the early twentieth century it was typical to invoke benefits to the group, or the species, as explanations of perceived adaptations. Among the most famous examples are V. C. Edwards's explanations of population regulation, such as through individual restraint in reproduction, as

Figure 1.7: Several insects and vertebrates such as caterpillars (e.g., *Arsenura armida*, pictured) and frogs ([Santos et al., 2003]) have evolved to be unpalatable, and to signal this unpalatability to predators via conspicuous warning markings. As noted by [Fisher, 1930], evolution of such a trait under natural selection acting at the level of the individual is problematic, as unpalatable and conspicuous individuals will bear a higher probability of being detected and eaten by a predator who has not yet experienced their unpalatability. Photograph by Artour Anker, reproduced with permission of the photographer.

being due to adaptation for the benefit of the population [Wynne-Edwards, 1962]. In response, George C. Williams argued comprehensively for the need to ascribe adaptation to the lowest plausible level, and distinguish adaptations from inevitable side effects. Thus, considering evolution of deer in response to predation, he argued that natural selection acting on individuals would give rise, as a side effect, to a "herd of fleet deer"; however selection on the group, he argued, would not be sufficiently powerful to shape an adapted group, a "fleet herd of deer," in which individuals each occupied distinct behavioral roles designed, in their interaction, to reduce the group's overall predation risk [Williams, 1966]. While Williams explicitly recognized the potential for selection acting between groups, which he took to be groups of unrelated individuals, rather than families or other relatives, like Darwin he ascribed the ultimate adaptive power to natural selection acting on individuals.

While for unrelated groups Williams gave primacy to individual-level selection producing individual-level adaptation, he noted that this was insufficient to explain many apparent incidents of self-sacrifice, such as the example of sterility in social insect workers introduced above. Williams did recognize and tackle this problem [Williams, 1966], but the priority belongs to another researcher whose work he referred to.

1.4   ENTER HAMILTON

The problem of providing an evolutionary explanation of self-sacrifice by individuals, as it stood in the mid-twentieth century, can be described no better than in the first published[4] biological writings of William D. Hamilton:[5]

> It is generally accepted that the behavior characteristic of a species is just as much the product of evolution as the morphology. Yet the kinds of behavior which can be adequately explained by the classical mathematical theory of natural selection are limited. In particular this theory cannot account for any case where an animal behaves in such a way as to promote the advantage of other members of the species not its direct descendants at the expense of its own. The explanation usually given for such cases and for all others where selfish behavior seems moderated by concern for the interests of a group is that they are evolved by natural selection favoring the most stable and co-operative groups. But in view of the inevitable slowness of any evolution based on group selection compared to the simultaneous trends that can occur by selection of the classical kind, based on individual advantage, this explanation must be treated with reserve so long as it remains unsupported by mathematical models. [Hamilton, 1963]

Hamilton went on to outline the evolutionary explanation of altruism, or self-sacrifice to benefit others, as follows:

> Despite the principle of "survival of the fittest" the ultimate criterion that determines whether [gene for altruism] G will spread is not whether the behavior is to the benefit of the behaver but whether it is of benefit to the gene G.... With altruism this will happen only if

the affected individual is a relative of the altruist, therefore having an increased chance of carrying the gene, and if the advantage conferred is large enough compared to the disadvantage to offset the regression, or "dilution," of the altruist's genotype in the relative in question. [Hamilton, 1963]

In sketching his new theory, Hamilton proposed that selection would favor the altruism gene G whenever

$$K > \frac{1}{r}, \tag{1.1}$$

where K is the ratio of the advantage conferred on recipients to the cost to the behaver, and r is a measure of genetic relatedness. Equation (1.1) is the first form of what came to be known as Hamilton's rule, a summary statement of when natural selection favors self-sacrificing behavior. The following year, Hamilton presented the mathematical basis for his suggestions [Hamilton, 1964a, Hamilton, 1964b]. What Hamilton achieved in these three papers was, as he put it, "an extension of the classical theory" [Hamilton, 1963], showing that genes within individuals should value not only the reproductive success of their bearer, but also the reproductive success of other individuals within the population who might carry the same gene with some degree of certainty. The "extension of the classical theory" was to observe that what mattered for natural selection was not simply the direct reproductive success of an individual (their *direct fitness*), but also their indirect reproduction via relatives containing the same genes, whose reproduction their own behavior had impacted on (their *indirect fitness*). Hamilton labeled this extended fitness as *inclusive fitness*, and described it thus:

> Inclusive fitness may be imagined as the personal fitness which an individual actually expresses in its production of adult offspring as it becomes after it has been first stripped and then augmented in a certain way. It is stripped of all components which can be considered as due to the individual's social environment, leaving the fitness which he would express if not exposed to any of the harms or benefits of that environment. This quantity is then augmented by certain fractions of the quantities of harm and benefit which the individual himself causes to the fitnesses of his neighbors. The fractions in question are simply the coefficients of relationship appropriate to the neighbours whom he affects. [Hamilton, 1964a]

As a simple word equation,

$$\text{inclusive fitness} = \text{direct fitness} + \text{indirect fitness}, \qquad (1.2)$$

where "direct fitness" in the above equation is the classical Darwinian/Fisherian conception of the fitness that natural selection acts on. In conceptualizing inclusive fitness, Hamilton had thus proposed what was arguably the most important extension of evolutionary theory since 1859. There were hints of such ideas before 1963 and 1964, however. Hamilton himself identified many of these [Hamilton, 1963], the most obvious being the recognition that parental care of offspring involves self-sacrificial behavior on the part of the parent, yet is consistent with standard Darwinian evolution. Hamilton also referred to R. A. Fisher's explanation of the evolution of aposematism (figure 1.7), which considered benefits to siblings from the evolution of conspicuous markings indicating distastefulness, that could arise only in the case that the unpalatable bearer of the markings was consumed and therefore forfeited all future reproduction [Fisher, 1930]. Hamilton noted, however, that Fisher did not consider benefits to relatives more distant than siblings; J.B.S. Haldane on the other hand did consider likelihoods of more distant relatives, up to cousins, containing genes for behavior when he laid out the evolutionary logic for a decision over whether to jump into a river to save a drowning child [Haldane, 1955], as Hamilton acknowledged [Hamilton, 1963].[6] Yet despite these isolated and partial appreciations of how self-sacrificing behavior might evolve, the priority remains with Hamilton for conceptualizing the solution so generally, and for presenting the first mathematical formalization of the arguments [Hamilton, 1963, Hamilton, 1964a, Hamilton, 1964b].

1.5   MULTILEVEL SELECTION THEORY

Inclusive fitness is not the only way of conceptualizing selection acting on social behaviors. An alternative viewpoint seems to spring from the earlier, discredited, view of adaptations for the benefit of the group [Wynne-Edwards, 1962] in that it considers group-level fitness. However, this new version of group selection theory explicitly decomposes total selection acting on a population into between-group selection, and within-group selection [Price, 1972a]. For altruism, it follows that when

$$\text{between-group selection strength} > \text{within-group selection strength}, \qquad (1.3)$$

altruism will be favored overall in a population[7] [Hamilton, 1975, Wilson, 1975a]. This theory has been labeled *group selection* [Price, 1972a, Hamilton, 1975] and *trait-group selection* [Wilson, 1975a], but is increasingly described as *multilevel selection* [Okasha, 2006]. Multilevel selection will be used in this book to describe scenarios in which condition (1.3) predicts the evolution of social behavior; most models of "group selection" postulated in the literature are of this type [Okasha, 2006].

Multilevel selection cannot explain the evolution of group-level adaptation, as early group selectionists believed [Okasha, 2006]; the problem is that when within-group selection is present then individual-level adaptations will be favored over group-level adaptations [Gardner and Grafen, 2009, Okasha and Paternotte, 2012]. However, multilevel selection theory is increasingly seen as an alternative to inclusive fitness theory, and one that is more general in its applicability. The details of these claims are given in later chapters, as is the demonstration that the multilevel selection rule (condition (1.3)) and Hamilton's rule (condition (1.1)) can always be applied to the same social trait, and always predict the same direction of selection.

1.6 The Generality of Inclusive Fitness Theory

This chapter has provided a brief historical introduction to the problem of apparent design in biology, evolutionary explanations of this, and in particular, evolutionary explanations of individual behaviors that appear designed to benefit not the individual themselves, but other members of their species. For other historical reviews of evolutionary theory, particularly with respect to the evolution of social behavior and the role of design in biological thought, the interested reader should see [Foster, 2009, Gardner and Foster, 2008, Gardner, 2009].

Despite the elegant simplicity of inclusive fitness theory, both in extending Darwinian theory, and also in explaining the evolution of self-sacrificing behavior, it has attracted criticism from some quarters, even fifty years after its advent. One particular line of criticism holds that multilevel selection is a more general theory of social evolution than is inclusive fitness theory. These misunderstandings will be identified and addressed later in this book, and provide one of the main motivations for its writing. The overarching motivation of this book is, however, to demonstrate and celebrate the generality of inclusive fitness theory, with a focus on its fundamental evolutionary logic. The remainder of the book is laid out as follows. The first third of the book presents the basic mathematical theory of natural selection, and inclusive fitness theory. Chapters 2 and 3 present two complementary approaches to building mathematical models of natural selection: the replicator dynamics and the Price equation, respectively. The replicator dynamics provide a very simple model of selection, and are used in chapter 2 to illustrate the action of natural selection on various kinds of social behavior, including nonadditive behaviors, when interactions are between nonrelatives. The Price equation, introduced in chapter 3, allows very general statements about selection to be formulated, and is used in chapter 4 to formalize the logic of inclusive fitness theory and derive Hamilton's rule in its simplest form, as well as show its equivalence with the multilevel selection approach.

The second third of the book treats more complicated social scenarios, and shows how inclusive fitness theory deals with these. Chapter 5 explores the outcomes of selection when interactions are nonadditive and occur between relatives, and explains two main approaches to generalizing Hamilton's rule to deal with such interactions. Chapter 6 considers behaviors that are expressed conditional on the phenotype of others, including the classic "greenbeard" thought experiment proposed by Hamilton [Hamilton, 1964b], and introduces a further generalization of Hamilton's rule to deal with such behaviors. Chapter 7 shows how the multiple versions of Hamilton's rule give different evolutionary explanations for certain traits such as greenbeards, and also how one can translate between these different versions. Chapter 9 deals with the problem of correctly defining fitness costs and benefits in inclusive fitness theory, when competition occurs between offspring who are relatives.

The final chapters of the book deal with more philosophical issues in inclusive fitness theory, to do with explanation of the results of selection in ultimate causal terms. Chapter 8 considers which of the equivalent alternative partitions of fitness, including inclusive fitness, and group fitness, can be interpreted as being subject to selection in a meaningful way, and also reviews proposals for classifying different evolutionary processes involved in the selection of social behavior. Chapter 9 reviews the definition of evolutionary fitness, and shows how its misinterpretation explains many previous misunderstandings as to whether inclusive fitness theory always makes accurate predictions. Finally, chapter 10 reviews the limitations of the analyses presented in this book, directing the reader to additional mathematical techniques, as well as considering the empirical support for inclusive fitness theory, and more advanced topics in the field.

CHAPTER TWO

Models of Social Behavior

2.1 INTRODUCTION

As we saw in the previous chapter, many animal traits and behaviors are social, in that they affect the reproductive success not just of the animal performing the behavior, but also conspecifics. Mathematical theories based on classical natural selection, which acts on direct reproduction by individuals, are able to explain the evolution of traits that are for personal advantage, regardless of whether those traits help or hinder the reproduction of other members of the general population. This leaves the problem, however, of providing an evolutionary explanation of traits and behaviors that appear to be personally costly to the bearer, in reproductive terms, while having effects on conspecifics such as increasing their direct reproduction. In tackling this problem Hamilton [Hamilton, 1963, Hamilton, 1964a, Hamilton, 1964b] presented a mathematical theory extending classical selection, showing that it should act on both direct reproduction by an individual, and indirect reproduction by genetic relatives that may share the genes that underlie social traits. In this chapter we will introduce the basic tools that allow us to explain and develop the mathematical theory of natural selection as applied to social behavior; note that in the remainder of this book we shall refer exclusively to social behavior, even though some social traits may not have behavioral components (e.g., figure 1.7).

A theory of social evolution requires a formal, mathematical way of modeling the effects of traits and behaviors on the reproductive success of individuals, and conspecifics. In fact, the mathematical study of social behavior increased throughout the twentieth century, motivated by economics initially. John von Neumann and

Oskar Morgenstern defined the field of game theory with their seminal *Theory of Games and Economic Behaviour* [Von Neumann and Morgenstern, 1944].[1] This approach formalized games, such as chess, and strategies for them, within a mathematical framework; it also treated problems of economic behavior as games susceptible to the same kind of analysis. Shortly afterwards, John Nash provided a formal definition of strategies that are best responses to themselves and that therefore cannot be displaced, and proved their existence [Nash, 1950, Nash, 1951]. The *Nash equilibrium* concept won its originator the Nobel Prize in Economics.[2] The economic view of games, which all social interactions can be conceptualized as, was naturally inspired by human behavior. Thus the analysis of the best course of action in a game rested on assumptions that the players were *rational*, could analyze the structure of the game and, assuming that other players would do the same, choose the action that should maximize their payoff from the game. Needless to say few, if any, of these assumptions could be considered safe in animals other than humans, and even their validity in humans has been extensively questioned.[3] Before too long, however, biologists realized that animals could still be shaped to behave optimally in social interactions, through the action of evolution. Although the idea of evolution arriving at an "unbeatable strategy" was traced, by John Maynard Smith [Maynard Smith, 1982], back to William D. Hamilton himself [Hamilton, 1967], it was Maynard Smith and George R. Price [Maynard Smith and Price, 1973] who first formalized this with their *evolutionarily stable strategy* (ESS) concept, thereby founding the field of evolutionary game theory [Maynard Smith, 1982].

Evolutionary game theory, like the original theory of games, models social interactions as games between a number of players. Here we begin by defining the necessary terminology, and introducing the simplest conceivable game. A *game* is an interaction between two or more *players*, each of whom has two or more *actions* available to them. We shall sometimes refer to the players engaged in a game as *partners*. Players choose actions according to a *strategy* which could be as simple as always choosing the same option, or as complicated as having a probability of playing each action, or choosing the action according to something about the other players, such as their actions on a previous play of the game. Each player receives a *payoff* from the game according to their choice of action and those of all the other players, with larger payoffs usually assumed to be more valuable to the player in some way. These are the minimum ingredients for a social interaction, or game; there should be at least two players, each should have a choice of actions, and the outcome of the game for each player should depend on what they and the other players choose to do. We can see that the simplest possible games then, will be those involving only two players, with only two choices each. These are typically referred to as the 2 × 2 *games*. For simplicity we also consider games where the payoffs are the same from both players' perspectives, so that the games are *symmetric*. The 2 × 2 games were first systematically investigated by Anatol Rapoport and Melvin Guyer [Rapoport and Guyer, 1966], and the four most interesting[4] symmetric games identified [Rapoport, 1967].

2.2 THE DONATION GAME

In table 2.1 we see our first game. The table has rows and columns corresponding to the action chosen by the first player and the second, respectively. The payoff the first (row) player receives depends on both these actions, and is discovered by looking at the appropriate row and column and reading the value in the cell that these index. Since, as we assumed above, the game is symmetric, the payoff the second player receives is found by simply swapping the row action with the column action and again reading off the indexed cell; thus, from table 2.1, if player 1 chooses **I** and player 2 chooses **II** then the first player gets $-c$ and the second gets b.

	I	**II**
I	$b - c$	$-c$
II	b	0

Table 2.1: The additive two-person donation game. Roman numerals in column and row headings denote the actions available to each player. Payoffs shown are for the row player.

Table 2.1 describes what we shall refer to as the *donation game*. It can be understood in the following terms: each player has the choice of whether to donate (action **I**) or not (action **II**). If a player chooses to donate they pay a cost $c > 0$, and the other player receives a benefit $b > 0$; we assume that $b > c$. Costs and benefits are frequently assumed to represent changes in the fitness of the individual and their partner, where fitness has a very particular evolutionary meaning; there are situations in which payoffs from a game matrix such as table 2.1 do not correspond to evolutionary fitness, however. In chapter 9 we shall more carefully consider the definition of fitness, and cases when payoffs from a game matrix do not map directly onto it.

There are several things worth noting about the donation game. First, the game is *additive*. This means that the total payoff a player receives is the sum of the payoff due to their own action, and the payoff due to the other player's action. Hence if player 1 chooses **I** and so does player 2, each gets the sum of the effect on their payoff due to their own action and due to their partner's. Second, the best strategy in the game is always to choose **II**. This can be seen from table 2.1 since, regardless of the other player's action, a higher payoff for the row player is always available by choosing action **II** rather than action **I**; that is, since we assumed that b and c are both greater than 0, then $b > b - c$ (column **I**) and $0 > -c$ (column **II**). Yet by following this logic both players converge on choosing action **II** and hence receive zero payoff, while if both choose **I** then each receives $b - c > 0$. The donation game is thus an additive version of the famous *prisoner's dilemma* of game theory [Rapoport, 1967, Axelrod, 1984], and below we shall show how evolutionary pressures can also result in the same dilemma that logical analysis of the payoff matrix leads to. The third, and final thing to note about table 2.1 is that when both players choose **II** then each receives zero payoff; this could be thought of as the baseline payoff that individuals have when they do not behave socially by donating, and there is no particular reason it should be 0 except for mathematical convenience. Of course individuals might receive payoffs due to causes other than the social interactions captured in the payoff matrix of the donation game; this nonzero baseline payoff would then be added to every cell in table 2.1 but, crucially, making such a payoff transformation never changes the equilibrium points of a game [Weibull, 1995]; hence it is simpler to leave baseline asocial payoff out, and describe the payoff matrix using only three explicit payoffs instead of four. This makes intuitive sense since, when we analyze selection on social behavior, what matters are the *changes* in payoffs that result from behavior.

2.2.1 Replicator Dynamics of the Donation Game

To show mathematically that the donation game with $b > c > 0$ favors players that choose not to donate (action **II**), we could undertake an equilibrium analysis. The introduction above was deliberately vague, however, on the mathematical definitions of Nash equilibria and evolutionarily stable strategies, since this book's focus is on understanding some important aspects of evolution, rather than on analyzing the equilibria of games. Many texts exist on this latter topic; for evolutionary game theory approaches in particular, see, for example [Weibull, 1995, Samuelson, 1998]. To understand the games considered in this book we will instead adapt a simple mathematical description of natural selection, the *replicator dynamics* [Taylor and Jonker, 1978]. The replicator dynamics is a staple tool of mathematical biology and shows how, under the assumption that strategies are faithfully passed on from parents to offspring and that payoffs in a game translate to production of offspring, the frequencies of different strategies in a population change over time. Assuming the simple case of two strategies, one being "always choose action **I**" and the other being "always choose action **II**," we let f be the relative frequency of strategy **I**. Since f must be between 0 and 1 inclusive, the frequency of **II** players must be $1 - f$. Then, given the payoffs of the donation game as shown in table 2.1, the replicator dynamics show that the change in relative frequency f is given by the equation[5]

$$\underbrace{\frac{df}{dt}}_{\text{change in } f \text{ over time}} = f\left(\overbrace{-c + fb}^{\text{payoff for strategy I}} - \left(\overbrace{\underbrace{f(-c + fb)}_{\text{strategy I payoff}} + (1 - f)\overbrace{fb}^{\text{strategy II payoff}}}^{\text{average payoff in population}} \right) \right).$$

$$(2.1)$$

Equation (2.1) is an ordinary differential equation, but those unfamiliar with them need not worry because (i) there are hardly any more examples in this book, (ii) the equation can be intuitively understood as saying that whether a strategy's relative frequency increases or decreases in a population depends on whether its average payoff is above or below the population average payoff, and (iii) this kind of equation can be graphically plotted very easily to see how it behaves for given parameters. In figure 2.1 we do this for arbitrary values $b > c > 0$, by representing the relative frequency f in a population at a given time as being a point on a line between $f = 0$ and $f = 1$, then for any point on the line, drawing an arrow pointing to the right if equation (2.1) is positive (f is increasing in frequency) and pointing to the left if equation (2.1) is negative (f is decreasing in frequency). In the case of figure 2.1 we see that the arrow points only to the left, indicating that players that always choose action **I** are driven to frequency 0 and remain there, regardless of their initial relative frequency.[6] The point $f = 0$ is thus a *fixed point* of the replicator dynamics, at which no further evolutionary change occurs due to selection. Because the population will return to the $f = 0$ equilibrium if disturbed from it, we refer more specifically to $f = 0$ as an *asymptotically stable* fixed point. Since $f = 0$ attracts nearby population states towards it, it is also referred to as

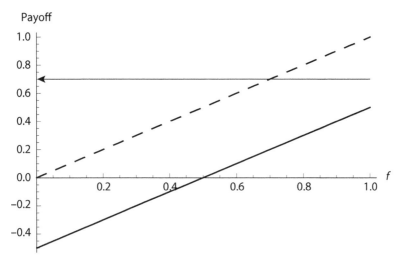

Figure 2.1: Visualization of selection pressures in the additive donation game. Solid and dashed lines show average payoffs to action **I** and action **II** players respectively, as the frequency of type **I** players, f, varies on the x-axis. Higher payoffs correspond to increased success under the replicator dynamics, and the arrow shows how f evolves under the replicator dynamics of equation (2.1), towards an attractor at $f = 0$. Parameters are $b = 1$, $c = 1/2$.

an *attractor* of the replicator dynamics; further, because f moves towards 0 from (almost) all points on the line, we refer to $f = 0$ as a *global attractor*. Figure 2.1 shows only that $f = 0$ is a global attractor for the particular values of $b > c > 0$ chosen; the beauty of a mathematical approach to analyzing selection in a game is that we can show the same outcome[7] for *all* values of $c > 0$. Thus, whenever costs of donation exist, then regardless of the size of benefits that they give rise to, selection always favors nondonation when individuals meet at random in a population.[8]

The replicator dynamics corresponds to the very simplest population genetical models, analyzing selection at a single haploid locus in an infinite population and hence ignoring other evolutionary processes and phenomena such as drift, mutation, gene transfer, linkage, and epistasis. They also relate to the equilibrium analysis approaches of classical and evolutionary game theory; Nash equilibria and asymptotically stable fixed points of the replicator dynamics are identical, whereas evolutionarily stable strategies correspond to attractors under the replicator dynamics, but not necessarily vice versa [Weibull, 1995]. The power of the replicator dynamics thus rests on their generality, their link with existing theories, and their simplicity. This generality and simplicity is based on an assumption, labeled the *phenotypic gambit* by Alan Grafen, who described it thus:

> The phenotypic gambit is to examine the evolutionary basis of a character as if the very simplest genetic system controlled it: as if there were a haploid locus at which each distinct strategy was represented by a distinct allele, as if the payoffs rule gave the number of offspring for each allele, and as if enough mutation occurred to allow each strategy the chance to invade. [Grafen, 1984]

Grafen discusses the nuances of applying the phenotypic gambit to analyzing an aspect of a real biological phenotype. Our aim in applying the replicator dynamics in this book is different; it is to answer the hypothetical question of what selective pressures act on different social behaviors in different kinds of social interactions, and examine the logic of these pressures. Throughout the book, we shall consider highly stereotypical and abstracted interactions and behavior, of the kind that can be captured in payoff matrices such as that of table 2.1. The expectation is that any abstract theory developed will then be applicable to real biological instances of social behavior, and any deviations from the predictions of the theory will be addressable by refinement of the theory to the particular details of the biological instance in question. As we shall see later in the book, this is indeed what has happened in the field of social evolution theory; the predictions of the original simple models have been abundantly confirmed by experiment, and deviations from those predictions have been explained by refinement of the theory.

2.3   THE NONADDITIVE DONATION GAME

In the previous section we were introduced to our first game: the additive donation game (table 2.1). As discussed in that section, social interactions in such games are additive in that an individual's payoff is the sum of the payoff arising from their own action and that arising from their partner's. Yet in biological systems, perfect linearity is likely to be the exception, rather than the rule. In this section we relax the assumption of additivity by introducing the nonadditive donation game, shown in table 2.2. Here an additional parameter d is added capturing the deviation from additivity that occurs when both individuals choose action **I**. This parameter can be positive or negative; crucially, changing d can radically alter the dynamics of the game as we shall see below.

	I	II
I	$b - c + d$	$-c$
II	b	0

Table 2.2: The nonadditive two-person donation game. Roman numerals in column and row headings denote the actions available to each player. Payoffs shown are for the row player.

2.3.1 Replicator Dynamics of the Nonadditive Donation Game

The replicator dynamics for the nonadditive donation game of table 2.2 is arrived at as for the additive donation game,[9] giving

$$\frac{df}{dt} = f\left(-c + f(b + d) - \left(f(-c + f(b + d)) + (1 - f)fb\right)\right). \quad (2.2)$$

This time it can be shown[10] that an equilibrium exists at $f = c/d$. Clearly if $0 < c/d < 1$ then this equilibrium is one that the population can actually occupy. Equally clearly, when $c > 0$ then this can be the case only if $d > 0$. We examine what happens under the replicator dynamics for this scenario below.

2.3.1.1 Positively Nonadditive Donation Games

When the parameter d in table 2.2 is positive, if both players choose action **I** then each donates b units of payoff to the other, but each also receives a *further d* payoff units. The game is thus *positively nonadditive*. Additivity is a very special case, so as already mentioned, additive biological interactions should be the exception rather than the rule. In general, we can refer to interactions that deviate from additivity as *synergistic* interactions[11] [Queller, 1984, Queller, 1985]. As discussed above, if donation is costly ($c > 0$) then if the positive nonadditive effect when two action **I** players interact exceeds the cost of donation ($d > c$), a *mixed-population* fixed point exists under the replicator dynamics. In fact it can be shown[12] that for $d > c > 0$ this fixed point is always unstable in that once the population is perturbed away from the equilibrium, it heads further away, much like a ball balanced on the top of a hill rolls down it when pushed. The effect of this is that a threshold population frequency exists, which donators must surpass if they are to become fully established in a population, as shown in figure 2.2.

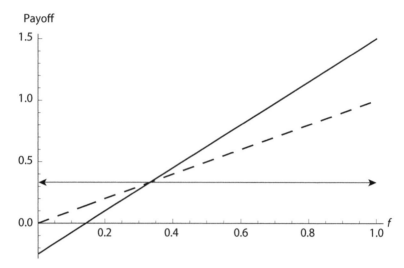

Figure 2.2: Visualization of selection pressures in the nonadditive donation game. Solid and dashed lines show payoffs to action **I** and action **II** players respectively, as the frequency of type **I** players, f, varies on the x-axis. Higher payoffs correspond to increased success under the replicator dynamics, and the arrows show how f evolves under the replicator dynamics of equation (2.2), away from a mixed-population fixed point at $f = c/d$. Parameters are $b = 1$, $c = 1/4, d = 3/4$.

2.3.1.2 Negatively Nonadditive Donation Games

Interactions between individuals can also be subject to *diminishing returns*, or *negatively nonadditive* ($d < 0$ in table 2.2). We saw above (section 2.3.1) that for nonadditive donation games there is only one possible mixed-population equilibrium, and this can be reached by a population only if interactions are positively nonadditive ($d > 0$). We also saw, when analyzing selection on additive games (section 2.2.1), that the magnitude of benefit b gained by recipients of donation is irrelevant in determining whether donation will be positively selected; all that matters is the sign of the cost parameter c, suggesting that as a costly behavior ($c > 0$), donation will experience negative selection at all population frequencies. In fact, we can show directly that costly donation always experiences negative selection when interactions are negatively nonadditive.[13] It should be emphasized, however, that this result is for interactions in fully mixed populations; in chapter 5 we will see what happens when populations are structured such that individuals tend to interact with similar partners.

2.4   OTHER SOCIAL INTERACTIONS

Although costly donation, often referred to as *altruism*, is almost certainly the most studied type of social behavior, other possible behavioral interactions exist. These different types of behavior are arrived at by considering all the possible combinations of effect on the behaving individual, and effect of the behavior on other individuals. Effects can be positive or negative.[14] Table 2.3 enumerates the possible behaviors and classifies them; in the following subsections, we explore the possible evolutionary outcomes of selection acting on such behaviors in unstructured populations.

Effect on own fitness	Effect on others' fitness	Classification
Positive	Positive	Mutual benefit / cooperation
Positive	Negative	Selfishness
Negative	Positive	Altruism
Negative	Negative	Spite

Table 2.3: Social behaviors can be classified according to their effect on the behaving individual's fitness, and on others' fitness. When fitness effects are additive, then in an unstructured population, behaviors experience positive or negative selection solely according to whether their effects on the behaving individual are positive or negative, as shown in this chapter. When behaviors interact to produce nonadditive fitness effects on individuals, however, other evolutionary outcomes are possible as shown in table 2.4. Classification labels are as proposed in [West et al., 2007b] with the exception of "cooperation," which West et al. use to denote both altruistic and mutually beneficial behaviors, but which is used in this book to refer only to mutual benefit.

2.4.1   Mutual Benefit

Mutual benefit arises when a behavior increases a behaving individual's direct fitness, but also increases the fitness of others at the same time. Such behavior is often also referred to as *cooperation*, to distinguish it from costly altruism.[15] Since mutual benefit increases direct fitness, it experiences positive selection. Even if the fitness benefit to others is greater than it is to the individual exhibiting the mutual-benefit-producing cooperative behavior, then despite the relative disadvantage of cooperators compared with noncooperators, cooperation still increases the direct

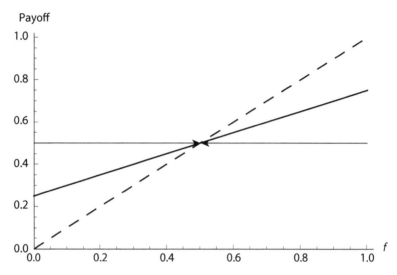

Figure 2.3: Visualization of selection pressures in the nonadditive cooperation game. Solid and dashed lines show average payoffs to action **I** and action **II** players respectively, as the frequency of type **I** players, f, varies on the x-axis. Higher payoffs correspond to increased success under the replicator dynamics, and the arrow shows how f evolves under the replicator dynamics of equation (2.2), such that there is a stable mixed-population equilibrium at $f = c/d$. Parameters are $b = 1, c = -1/4, d = -0.5$.

fitness of individuals and therefore experiences positive selection.[16] If, however, there are diminishing returns from cooperation, so fitness effects are negatively nonadditive ($d < 0$), then a stable *mixed equilibrium* of cooperators and noncooperators can exist,[17] as illustrated in figure 2.3. The occurrence of cooperation in the natural world can thus be explained by standard natural selection theory acting on individuals' direct fitness.

2.4.2 Selfishness

Selfish behavior increases the direct fitness of the individual expressing it, but reduces the fitness of others, as summarized in table 2.3. In an unstructured population the benefit to the individual is key, while the effects on others' fitness are irrelevant, hence selfishness experiences positive selection as shown in this chapter both for interactions between pairs of individuals and interactions within larger groups. As summarized above for cooperation, and in table 2.4, when a behavior is individually beneficial then negatively nonadditive interactions between individuals can lead to a stable population equilibrium containing a mixture of social (selfish) and asocial (nonselfish) individuals.

Effect on own fitness	Additivity	Economic game	Outcomes
Negative	Zero *or* negative	Prisoner's dilemma [Tucker, 1950, Rapoport, 1967] / public goods game [Olson, 1965]	Asocial type dominates
Negative	Positive	Stag hunt [Skyrms, 2004]	Asocial type dominates *or* bistability if positive nonadditivity large enough
Positive	Negative	Chicken [Rapoport, 1967] / hawk–dove [Maynard Smith, 1982] / snowdrift [Sugden, 1986]	Social type dominates *or* mixed equilibrium if deviation from additivity large enough
Positive	Zero *or* positive	—	Social type dominates
Negative	Step / sigmoidal (figure 2.6)	Volunteer's dilemma [Dieckmann, 1985]	Mixed equilibrium [Archetti and Scheuring, 2011]

Table 2.4: Different effects on fitness associated with a social behavior correspond to different economic games, and in unstructured populations result in different evolutionary outcomes. These outcomes are fixation of the social behavior, fixation of the asocial behavior, bistability so that each behavior can fixate if it achieves high enough frequency, and mixed equilibrium in which a stable population contains both social and asocial types. The economic game labels shown here assume that social behavior by an individual results in positive fitness benefits for partners the individual interacts with. However, in unstructured populations, as considered in this chapter, the effect on other individuals' direct fitness is irrelevant. For pairwise interactions, the direction of selection is independent of effects on others' fitness (see notes 7 and 10); hence the same evolutionary outcomes occur for altruistic and spiteful behaviors. Similarly, for nonadditive interactions between more than two individuals, the possible evolutionary outcomes depend only on the net effect of social behavior on a focal individual's direct fitness, and not on the effects on other individuals' fitness (see notes 19 and 20).

2.4.3 Spite

Spite describes behaviors that reduce the direct fitness of others, as in selfishness above, but at a cost; spiteful behaviors reduce the direct fitness of the behaving individual, just as altruistic donation reduces the direct fitness of a donating individual. Just as for altruism, therefore, the reduced direct fitness of spiteful individuals means they experience negative selection (see note 7 and section 2.5) in unstructured populations. However, just as for altruism again, positively nonadditive behavioral interactions between spiteful individuals can result in a threshold frequency, above which spite goes to fixation in the population. This is shown in tables 2.3 and 2.4, for both pairwise interactions (see note 12) and in interactions within groups of more than two individuals (as described in the following section).

2.5 PUBLIC GOODS GAMES

Additive and nonadditive social behaviors have so far been considered in terms of interactions between two individuals, as in tables 2.1 and 2.2. Yet many biological interactions are not of this type. To take one obvious example, social behaviors of microbes, as in figure 1.6, involve many more than two individuals. Clearly therefore we need a model of social interaction involving multiple individuals. Economists have studied a version of such games, in which a social behavior is individually costly but leads to a group benefit, under the label of *public goods games* [Samuelson, 1954, Olson, 1965]. Typically public goods games are considered in their additive form, in which benefits to the group grow linearly with the number of group members performing the social behavior. Such interactions have been generalized by theoretical biologists, however, in a pleasingly simple model of potentially nonadditive public goods games [Hauert et al., 2006]. Cristoph Hauert and colleagues[18] present a model under which each additional prosocial group member either contributes a constant additional benefit to the group (in the case of additive interactions), or an additional benefit that is multiplicatively reduced relative to the previous member's contribution (negative nonadditivity) or multiplicatively increased relative to the previous member's contribution.[19] As summarized in table 2.4, in unstructured populations the possible evolutionary outcomes under different net costs to social individuals, and different forms of nonadditivity, are exactly the same for larger group sizes as they are when interactions occur within pairs of individuals. In particular, when social behavior provides a net fitness benefit to the individual and interactions are negatively nonadditive, then stable populations with a mixture of

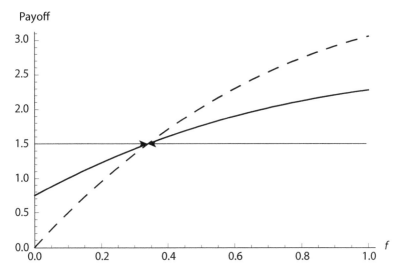

Figure 2.4: Visualization of selection pressures in the negatively nonadditive group cooperation game. Solid and dashed lines show average payoffs to action **I** and action **II** players respectively, as the frequency of type **I** players, f, varies on the x-axis. Higher payoffs correspond to increased success under the replicator dynamics, and the arrow shows how f evolves under the replicator dynamics, such that there is a stable mixed-population equilibrium. Since the net effect of contributing to the group good on personal direct fitness is positive (i.e., $b/n > c$), then leaving aside nonadditivity, the behavior is mutually beneficial rather than altruistic (table 2.3). Parameters are $b = 7, c = 1, \delta = 1/2, n = 4$.

social and asocial individuals are possible (figure 2.4), and when social behavior is individually costly and interactions are positively nonadditive, then either social or asocial populations can be stable (figure 2.5). The possible evolutionary outcomes are unchanged regardless of whether the social behavior increases or decreases others' direct fitness, so selfishness and spite can also be modeled; all that matters for the direction of selection on the social behavior is whether it has a positive net effect on the individual expressing it.[20]

2.6   THRESHOLD PUBLIC GOODS GAMES

In all of the social interactions modeled above, from additive interactions within pairs of individuals through to positively or negatively nonadditive interactions in group sizes larger than two, under no circumstances are social individuals that reduce their own direct fitness in order to improve that of others able to invade

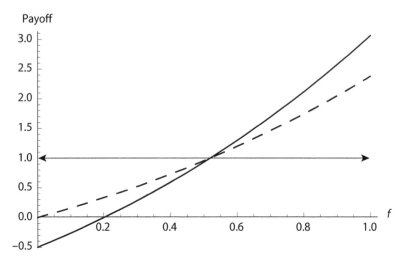

Figure 2.5: Visualization of selection pressures in the positively nonadditive public goods game. Solid and dashed lines show average payoffs to action **I** and action **II** players respectively, as the frequency of type **I** players, f, varies on the x-axis. Higher payoffs correspond to increased success under the replicator dynamics, and the arrow shows how f evolves under the replicator dynamics, such that exclusively social, and exclusively asocial, populations are both stable. Since the net effect of contributing to the group good on personal direct fitness is negative, then leaving aside nonadditivity (i.e., $b/n < c$), the behavior is altruistic (table 2.3). Parameters are $b = 2, c = 1, \delta = 3/2, n = 4$.

unstructured populations; the best situation for favoring the evolution of social behavior is when interactions have a sufficiently large positively nonadditive component, in which case social behavior can go to fixation if enough social individuals arise in the population. This is important for two reasons. First, as discussed in chapter 1 we are interested in explaining the evolution of self-sacrificing social behavior, since it seems to present a challenge for classical Darwinian natural selection. Second, positive nonadditivity does not resolve the problem because a threshold frequency of social individuals must be surpassed for them to take over a population, whereas the first occurrence of social behavior is likely to be as a single mutant.

Marco Archetti and Istvtán Scheuring have, however, highlighted a class of social interactions in which individually costly behavior can experience positive selection from its initial emergence in a population, and result in a stable mixed-population equilibrium in which social and asocial individuals coexist [Archetti and Scheuring, 2011]. Archetti and Scheuring's starting point is the classic *volunteer's dilemma* of social science [Dieckmann, 1985], in which

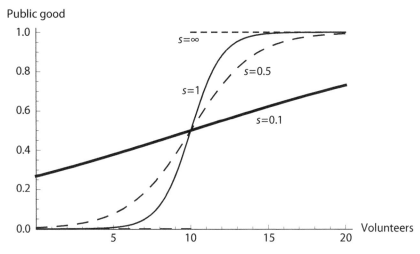

Figure 2.6: Group payoffs in the threshold public goods game. [Archetti and Scheuring, 2011] propose a group payoff function (2.19), as a function of the number of prosocial volunteers within a group, that interpolates a pure volunteer's dilemma ($s = \infty$), in which a public good is produced only if enough volunteers contribute to it, and a standard additive public goods game (as s approaches 0) in which group benefit is a linear function of the number of prosocial group members. Archetti and Scheuring show that, with random formation of groups, volunteer's dilemmas ($s > 0$) result in a stable mixed-population equilibrium consisting of prosocial and asocial individuals (equation (2.3)). Figure adapted from [Archetti and Scheuring, 2011].

individuals choose whether to contribute to the production of a public good which, if produced, will be sufficiently valuable to them as to outweigh the cost of their contribution. However, the public good is produced and shared only if a sufficient number of group members volunteer. Archetti and Scheuring propose a generalized volunteer's dilemma group benefit, as a function of the number of group members contributing and the threshold contribution required, that interpolates the pure volunteer's dilemma and the additive donation game as illustrated in figure 2.6.[21] Archetti and Scheuring construct a model of random formation of groups, and show that when the public good production is sigmoidal or step shaped, a stable-intermediate frequency of prosocial volunteers occurs, at approximately[22]

$$f^* = \frac{m - 1}{n - 1}, \tag{2.3}$$

where m is the number of volunteers required to produce the public good, and n is the group size as before. When the volunteer's dilemma is generalized so that additional volunteers always increase the size of the group good (see note 21 and figure 2.6), then m is the threshold at which further volunteers change from having a positively nonadditive effect on the size of the group benefit, to having a negatively nonadditive, diminishing returns, effect on the group benefit's size. Thus, the bistability result of [Archetti and Scheuring, 2011] rests on having both kinds of nonadditivity within the same payoff function: positive nonadditivity for lower frequencies of prosocial group members, and negative nonadditivity for higher frequencies. Selection leads to a stable frequency of social individuals within the population purely through the benefits to an individual's direct fitness that result from volunteering when such volunteers are at low enough population frequencies [Archetti and Scheuring, 2011].

2.7 INTERACTIONS IN STRUCTURED POPULATIONS

This chapter has highlighted how, in unstructured populations, the effects of behaviors on the fitness of others are irrelevant; the effect of a behavior on its bearers' direct fitness predicts whether they experience positive selection. Behaviors will experience positive or negative selection according to whether they respectively increase or decrease an individual's fitness. The only exception to this is when behaviors in different individuals interact nonadditively; in this case, nonadditivity can lead to either a stable mixed equilibrium of social and asocial types in the population, or result in a threshold frequency that the social type must surmount in order to go to fixation in the population. Fundamentally, however, the frequency-dependent direct fitness effect of a behavior still predicts the direction of selection on it.

In the coming chapters we will see how, when populations are structured such that interactions occur between similar individuals, the effect of behaviors on others' fitness becomes crucial in determining the response to selection. Before we do so, however, in the next chapter we must first introduce another general representation of selection and evolution: the Price equation.

2.8 SUMMARY

This chapter has presented the basic classification of social behaviors according to whether they directly benefit the individual performing them, and whether they benefit social partners. In freely mixing populations, behaviors that have a negative effect on the performing individual generally will not spread under natural selection, whereas those having a positive direct effect will. However, there are important exceptions to this. In particular, positively nonadditive but personally costly behavior can lead to a threshold frequency of prosocial individuals which, once passed in a population, will lead under natural selection to a completely social population. Personally beneficial behavior that also benefits others, on the other hand, will spread to an intermediate frequency in the population only when those benefits are negatively nonadditive. These results hold for pairs and for groups of more than two individuals. Finally, personally costly social behavior can spread to an intermediate frequency in populations when it is positively nonadditive at low frequencies, and negatively nonadditive at high frequencies.

CHAPTER THREE

The Price Equation

3.1 A GENERAL DESCRIPTION OF SELECTION

In the previous chapter we were introduced to one simple and general model of selection: the replicator dynamics. In this chapter, we examine an even more general description of selection: the *Price equation*. Developed by George Price[1] in the late 1960s [Price, 1970], the Price equation is a fully general description of selection, that can be applied to the change of any quantity under any selective regime.[2] The equation is thus not limited to considering simple haploid single-locus traits, unlike the replicator dynamics, and indeed it is not even limited to considering evolutionary selection, as we shall see below. This generality, however, tends to come at a cost. As applied to many typical problems, the Price equation describes only the change in the average of some quantity in response to selection at a single point in time; this is in contrast to the equations of the replicator dynamics, which can be used to calculate a population's trajectory from a present state to an ultimate evolutionary attractor. The Price equation thus provides us with an instantaneous description of selection in action. The great advantage of this approach is that its generality and, as we shall see, its simplicity, provide a useful conceptual tool for understanding selective processes such as natural selection.

We start by describing the general Price equation. Since this is a book on social evolution, we shall approach the equation using the terminology of populations undergoing selection, but we shall also see a nonbiological example shortly. Price considered a "parent" population, each member of which has some *genic value*[3] associated with them, and each of which produce zero or more offspring having their own genic value [Price, 1970]. Price described these genic values as if they were "doses" of alleles within individuals, so for a single-locus trait, a diploid individual could have a value of 0, 1/2, or 1, according to whether they have 0, 1, or 2 copies of the relevant allele, respectively. While Price described individuals' values in terms of gene frequencies, he emphasized that any linear combination of genes at different loci could still be handled by the theory;[4] this has an important benefit that we will return to below.

To present the equation, we represent parent individuals' genic values as g_i, where i is the ith individual in the parent population. The ith parent has w_i offspring, where $w_i \geq 0$; the average genic value of these offspring is then denoted g_i'. We let Δg_i be the change between the genic value of parent i and the average genic value of its offspring. Then, as [Price, 1970] showed, the change in the average genic value of the population from the parental generation to the offspring generation is[5]

$$\overline{w} \underbrace{(\overline{g'} - \overline{g})}_{\text{genetic change}} = \underbrace{\text{Cov}(g, w)}_{\text{change due to selection}} + \underbrace{\overline{\Delta g},}_{\text{change due to transmission}} \tag{3.1}$$

where $\overline{w}$, for example, is the average of all the parents' individual offspring numbers,[6] and $\text{Cov}(g, w)$ is the statistical covariance between the parents' genic values, and their offspring numbers.[7] It is important to note that the averages relating to the offspring population, that is, $\overline{g'}$ and $\overline{\Delta g}$, are averages of the average values of parents' offspring, weighted by the frequency of those offspring.[8] The Price equation divides the total genetic change, typically denoted[9] Δg, into the part of that change due to selection acting on parents according to their genic value, and the part due to deviations between parents' genic values and their offspring's genic values.[10] It can often be shown, or assumed, that there is no systematic deviation between parent and offspring values in which case the second, transmission, term is zero on average[11] and evolutionary change is entirely due to response to selection, so

$$\overline{w} \Delta g = \text{Cov}(g, w). \tag{3.2}$$

Later in the book, however, we shall see how interesting evolutionary problems can be analyzed using the Price equation, for which the transmission term is systematically nonzero.

Although presented in genetical terms, the Price equation is general enough to describe any kind of response to selection, as we see immediately below.

3.1.1 Student IQs and Selection

In presenting his approach to describing selection, Price emphasized its generality by means of a simple example, analyzing change in the average IQ of students taking a university course and sitting an exam. As he put it,

> If students' expectations of passing a certain course vary with IQ, and if student IQs do not change appreciably during the course, then equation [(3.2)] (with its variables suitably redefined) will give the difference in mean IQ between students entering the course and those completing it (and equation [(3.1)] will apply if IQs do change during the course). [Price, 1970]

The value of this generality will become apparent later in this chapter.

3.2 GENETIC SELECTION

Although Price emphasized the generality of his covariance approach to selection, his primary interest was in its application to genetic evolution, writing that "gene frequency change is the basic event in biological evolution" [Price, 1970]. Price saw the benefit of his approach as having "great transparency, making it useful as a tool in qualitative evolutionary reasoning." Central to this transparency is the Price equation's link with simple data analysis using linear regression.

3.2.1 Selection as Regression

Price noted that a covariance, as appears in his equation, is proportional to a linear regression coefficient calculated using the least-squares method.[12] This led him to propose a regression coefficient form of his equation (see note 12), and to observe,

> Equation [(3.1)] in its regression coefficient form can be visualized in terms of a linear regression line fitted to a scatter diagram of w against g. (A linear regression line is the best construction in terms of the population effect Δg even if it gives a poor fit in terms of individual points.) Since the regression line has slope β_{wg}, gene frequency change due to selection is exactly proportional to the slope. Therefore, at any step in constructing hypotheses about evolution

> through natural selection — for example, about why human canines do not protrude, why deer antlers are annually shed and renewed, why parrots mimic, why dolphins play — one can visualize such a diagram and consider whether the slope really would be appreciably non-zero under the assumptions of the theory. If there is no slope, then there is no frequency change except by Δg_i [transmission] effects, and the hypothesis is probably wrong. [Price, 1970]

So, Price conceptualized the application of his equation to a question of whether selection occurs, in terms of a simple experiment. In this experiment one observes the genic values of parents for some trait of interest, and their numbers of offspring, then performs a linear regression analysis; a positive regression slope indicates that the trait is under positive selection (individuals with "more" genes for the trait leave more offspring, on average), whereas a negative slope indicates negative selection on the trait (individuals with "fewer" genes for the trait leave fewer offspring, on average). Figure 3.1 presents just such an analysis, for antler size in red deer[13] [Kruuk et al., 2002]; a similar analysis for a bacterial trait is shown in figure 1.5.

3.2.2 Models and Statistics

The Price equation has been presented above as a descriptive tool, to reason about data from experiments. Using this tool, one can measure individuals' genic values for a trait (or functions of them),[14] and offspring production, then ask whether the two positively or negatively covary, and to what extent.[15] This book, however, is about social evolution theory, and its conceptual underpinnings. The Price equation has enabled substantial clarifications of the fundamentals of social evolution theory, through facilitating "qualitative reasoning" about simple models of social evolution. Thus we need to translate between a procedure applicable to statistical data, and one applicable to models of evolution.

To arrive at a version of the Price equation applicable to probabilistic models of evolutionary processes, we play a simple mental trick. Imagine we expand our sample of individuals, for which we can calculate a particular instance of the Price equation, to be infinitely large and include all the possible individuals that could ever exist, represented with a frequency proportional to the probability with which they can exist; this is the *statistical population*.[16] Since this may seem a little abstract, consider coin tosses from a two-sided coin that has a particular probability p of coming up heads. A particular set of toss outcomes, such as "heads, heads, tails,

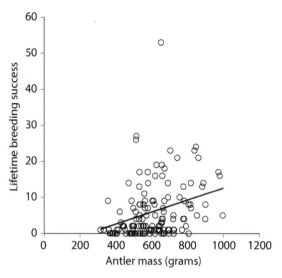

Figure 3.1: Lifetime breeding success and antler mass in red deer. Antler mass is a phenotypic trait, but assuming it to be under simple genetic control, so more "doses" of an "allele for antlers" result in larger antlers on average, then the Price-equation approach would be to regress individuals' lifetime breeding success (i.e., offspring number w_i) on antler allele dose (i.e., genic value g_i), and determine whether the slope of a linear regression is positive or negative. If positive, then alleles for antler mass would be under positive selection, and if negative then those alleles would be under negative selection. In this case, the slope of a linear regression is positive (but despite being heritable and undergoing selection, this particular trait in this particular population is not evolving, as discussed in the penultimate section of this chapter). Figure adapted from [Kruuk et al., 2002], with original data provided by Loeske Kruuk.

heads," is a statistical sample for which, if we assign values to the different outcomes such as 1 for heads and 0 for tails, we can compute a mean, a variance, and so on. The statistical population this sample is drawn from is the set of all possible tosses of that coin, and this also has a mean (p), a variance ($p(1 - p)$), and so on. We can represent this statistical population of coin tosses as a *random variable*; let us call it C. Means of random variables are generally not called means, but rather *expectations*, and we will denote them $E(C)$, for example. Variances are still variances, so for our coin-toss random variable the variance would be denoted $Var(C)$. Expectations and variances of random variables can be thought of as averages and variances of statistical samples, when those samples become infinite.[17] We have lingered over the definition of random variables, and their expectation and (co)variance, because from this point onwards we shall encounter the Price equation only as applied to probabilistic models, constructed using random variables such as the genic value of focal individuals and their social partners. To denote random variables we shall follow the typical practice of representing them using capital letters. With this in mind, our version of the Price equation applicable to probabilistic models is[18]

$$\mathrm{E}(W)\Delta\mathrm{E}(G) = \mathrm{Cov}(G, W) + \mathrm{E}(W\Delta G). \qquad (3.3)$$

The question has been asked, when applying the Price equation, is one doing statistics, or modeling [van Veelen, 2005]? The answer, as we have just seen above, is that it can do both. When applied to a set of observations of a natural population, one gains an insight into whether selection is favoring a trait, in that particular population, at that particular point in time (e.g., figure 3.1). When applied to a probabilistic model, the Price equation gives the expected evolutionary change at a particular moment in time, as a function of the model parameters. It is this latter use that we shall put the Price equation to in this book, as was Price's intention (see note 5).

Frequently in our models we will know that transmission of traits is faithful on average, hence $\mathrm{E}(W\Delta G) = 0$. We can then also see that, since $\mathrm{E}(W)$ must always be positive, if all we care about is whether a trait receives positive selection, and we do not care how much selection the trait receives, then we need consider only a simplified Price equation and ask under what conditions

$$\mathrm{Cov}(G, W) > 0. \qquad (3.4)$$

This will be useful in later chapters for deriving classical results from inclusive fitness theory and multilevel selection theory.

3.3   ILLUSTRATIVE APPLICATIONS OF THE PRICE EQUATION

To cement our understanding of the Price equation, and appreciate the kind of qualitative evolutionary reasoning it can provide, we now see how it can be used to derive two classical results from population and quantitative genetics: Fisher's "fundamental theorem of natural selection" [Fisher, 1930] and the breeder's equation [Falconer and Mackay, 1996].

3.3.1   Fisher's Fundamental Theorem

Among the many contributions of R. A. Fisher, one of the principal architects of the modern synthetic theory of evolution (see chapter 1), is a theorem that long presented a puzzle for other evolutionary biologists. Referred to as the *fundamental theorem of natural selection* by Fisher himself, and often referred to as *Fisher's fundamental theorem*, it states,

> The rate of increase in fitness of any organism at any time is equal to its genetic variance in fitness at that time. [Fisher, 1930]

Fisher likened this to the second law of thermodynamics, that the entropy of an isolated system never decreases, and considered it to have "the supreme position among the biological sciences." While Fisher described the theorem as rigorous and easy to interpret, for many decades after its publication other theoreticians struggled to understand the result and rederive it with the generality that Fisher claimed.[19] George Price resolved the problem by noting that Fisher was not concerned with the total increase in fitness, but rather the fitness increase due to natural selection, independent of the fitness change due to "environmental" change. Interestingly, Price did not apply his covariance approach to Fisher's fundamental theorem, but Steve Frank and Montgomery Slatkin subsequently showed how this could be done along the following lines [Frank and Slatkin, 1992].

We start by recalling that the Price equation can be applied to describe evolutionary change in *anything*, hence we ask about the evolutionary change in fitness $\Delta E(W)$. Next we partition the total evolutionary change in fitness as

$$\Delta E(W) = \Delta_S E(W) + \Delta_E E(W), \qquad (3.5)$$

where $\Delta_S E(W)$ is the evolutionary change in fitness due to selection, and $\Delta_E E(W)$ is the evolutionary change in fitness due to environmental change.[20] If we are interested only in the change in fitness due to selection, and assuming offspring inherit their parents' fitnesses faithfully on average, then since $\Delta_S E(W)$ is defined by a version of the Price equation we can rewrite it as[21]

$$\Delta_S E(W) = \mathrm{Var}(W)/E(W), \qquad (3.6)$$

where $\mathrm{Var}(W)/E(W)$ is variance in fitness relative to the population mean. Equation (3.6) is Fisher's fundamental theorem, stating that at any point in time the change in fitness due to selection, and excluding environmental effects, equals the additive genetic variance in fitness at that time, where fitness is relative fitness, normalized by population average fitness. There is a peculiarity in the fundamental theorem, that the environmental effects captured in the $\Delta_E E(W)$ term also include

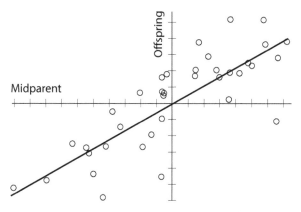

Figure 3.2: Heritability of wing length in *Drosophila*, estimated from experimental data as the slope of a linear regression of mean offspring length on mean parent length. The origin of the graph is the grand mean of all parental values against all offspring values, and axis intervals represent 0.01 mm. Figure redrawn from [Falconer and Mackay, 1996].

the genetic environment of the trait under consideration, in terms of frequencies of alleles at that trait locus as well as at others, and these are themselves the result of selection; this led some, such as George Price [Price, 1972b] to feel that the theorem was of limited importance,[22] whereas others such as Frank and Slatkin still consider it fundamental [Frank and Slatkin, 1992].[23]

3.3.2 The Breeder's Equation

We now see how the Price equation can be applied to derive one of the classical results of quantitative genetics: the breeder's equation. The breeder's equation was developed for practitioners of artificial selection, such as cattle breeders seeking to maximize milk yield, and states that the total evolutionary change in a quantitative trait is given by

$$R = sh^2,\tag{3.7}$$

where R is the *response of the trait to selection*, s is the *selection differential*, and h^2 is the *heritability* of the trait [Falconer and Mackay, 1996].[24] Heritability is the extent to which offspring's trait values correlate with those of their parents, as illustrated in figure 3.2 for *Drosophila*. Since the genotypic makeup of individuals is often not directly observed, and the genetic bases of traits are often not fully understood, the breeder's equation and quantitative genetics in general make use of an indirect measure of the genetic contribution underlying an individual's particular phenotype: the *breeding value*. Breeding value is assigned to individuals according to the mean value of their offspring that arise from random mating into a reference population [Falconer and Mackay, 1996]; breeding value is thus a linear function of an individual's genes, so R in equation (3.7) is the change in mean breeding value of a population, and this can therefore be studied using the Price equation.

In applying the Price equation this time, rather than considering change in gene frequency or average fitness as in the preceding sections, we will use it to describe the change in average breeding value in a population. However, since quantitative genetics is concerned with the evolution of quantitative phenotypic traits, such as weight, height, and so on, our analysis will introduce a phenotype that is under, at least partial, genetic control. To do this we follow the approach of David Queller [Queller, 1992b] in relating breeding value G to phenotype P. Queller's approach is to describe breeding value G in terms of its least-squares regression on phenotype P, so

$$G = \alpha_{GP} + \beta_{GP} P + \varepsilon_{GP}, \tag{3.8}$$

where, as before, β_{GP} is the slope of the least-squares regression of G on P, α_{GP} is the intercept of that regression, and ε_{GP} captures the residuals of the regression which arise when particular values of G fall off the line predicted by equation (3.8).[25] Since genotypes cause phenotypes, rather than the other way around, regressing something genetic on something phenotypic may seem counterintuitive. However, it makes perfect sense from the perspective of trying to predict what we typically cannot observe (genes underlying a particular trait) from what we can observe (phenotypic value for that trait).

Now, again assuming that on average offspring faithfully inherit traits from their parents, since breeding value is a linear function of allele frequencies we can use the Price equation to show that genetic change in the population under selection is[26]

$$\underbrace{\mathrm{Cov}\left(G, \frac{W}{\mathrm{E}(W)}\right)}_{\text{response to selection on trait}} = \overbrace{\mathrm{Cov}\left(P, \frac{W}{\mathrm{E}(W)}\right)}^{\text{selection differential}} \underbrace{\beta_{GP}}_{\text{heritability of trait}}, \tag{3.9}$$

which is the breeder's equation (3.7) with $R = \mathrm{Cov}(G, W/\mathrm{E}(W))$, $s = \mathrm{Cov}(P, W/\mathrm{E}(W))$, and $h^2 = \beta_{GP}$ [Queller, 1992b].[27] Thus the Price equation allows us to derive total evolutionary change as being the product of the relationships between phenotypes and fitness, and between phenotypes and genes.

3.4 IMPORTANT CAVEATS

Although the Price equation provides us with a simple and intuitive way to consider selection in general, and natural selection in particular, a number of caveats in its application should be noted.

The first caveat in applying the Price equation is the old statistical adage, "correlation is not causation." This can be illustrated very simply by noting that many traits are *pleiotropic*, having multiple effects, and selection acts on the ensemble of traits that an individual possesses. Thus, imagine for example an allele simultaneously coding for wings and lime green eyes. In this example, imagine selection favors flight and is neutral on eye color; hence we see a positive covariance between having wings and reproductive success, which the Price equation suggests we should interpret as evidence for the phenotypic trait being under positive selection. However, the same positive covariance also exists between green eyes and reproduction, which should lead us to conclude that it is also under positive selection, under the logic of the Price equation, even though it is selectively neutral. One solution to this can be to apply a multivariate version of the breeder's (Price) equation to detect such spurious correlations and correctly ascribe responses to selection (e.g., [Futuyma, 1998, Morrissey et al., 2010]).

Loeske Kruuk and colleagues discuss a similar problem in considering why antler size in red deer appears to be both heritable and under positive selection (figure 3.1), yet did not exhibit an evolutionary change over a 30-year study period [Kruuk et al., 2002]. In explaining this they note an "environmental" component to both antler size and reproductive success, in that the nutritional state of males facilitates both the growth of large antlers, and the winning of fights over access to mates.

The above examples illustrate a statistical problem to do with inferring causation from experimentally observed correlations. In probabilistic models the same caveat applies; however, since such models are formulated according to our own definitions, rather than being unknown experimental objects that we must interpret data from, we are able to equate correlation with causation when appropriate, taking proper account of the logical structure of the model we have built[28] [Okasha, 2006].

One must also note that individuals differ in their potential contributions to evolutionary change, through being members of different classes. Price himself noted the importance of accounting for sex differences in applying his equation, showing how it must be corrected to deal with alleles that reside only on the X chromosome, for example [Price, 1970]. As Fisher pointed out, one must also account for the varying reproductive potential of individuals, such as those of different age[29] [Fisher, 1930], sex, size, or any other aspect of individual state influencing reproductive potential [Houston and McNamara, 1999]. Fisher termed this *reproductive value*; a simple example of its importance can be illustrated by considering which parent has been more successful at making a genetic contribution to future generations: the parent of ten weak offspring that are unlikely to survive to reproductive age, or the parent of two healthy offspring with sufficient reserves to give them a good chance of surviving to raise offspring of their own. The Price equation can also be extended to deal with such class-structured populations [Taylor, 1990] (see also chapter 10).

Finally, one must be cautious in extrapolating the Price equation applied to describe change over a single generation, in order to predict longer-term evolutionary change. Even without differences in reproductive value among different classes of offspring, predicting long-term evolutionary change with the Price equation can be problematic. While the replicator dynamics of the previous chapter can be iterated to determine long-term evolutionary trajectories, and thus are *dynamically sufficient*, the Price equation as applied to many kinds of model is often not. This is a feature of the models considered rather than the Price equation per se [Frank, 1995], although authors sometimes confound the two (e.g., [Traulsen, 2010]; see also chapter 10). In terms of the Price equation, the problem is that after one generation the equation predicts only the mean value of something after selection (the first moment of its distribution), whereas applying the Price equation to the next generation, with offspring now acting as parents, requires knowledge of the variance (second moment) as well as the mean. Andy Gardner and colleagues term this the *moment closure problem* [Gardner et al., 2007b]. In fact Alan Grafen has shown that, given some very simple assumptions, a large class of models, particularly simple haploid population genetics models of the kind captured by the replicator dynamics, can be described by a multigenerational version of the Price equation[30] [Grafen, 2000]. Applied to a linear function of genic values, however, such as breeding value, the Price equation (or any other modeling formalism making use of the same level of abstraction [Frank, 1995, Frank, 2012]) cannot be applied recursively since gene frequencies, and thus linear functions of them such as breeding values, change under selection [Grafen, 2000].

3.5 SUMMARY

In this chapter we have shown how the Price equation has been useful in understanding the logic of evolution through natural (and artificial) selection. The evolutionary cases we have considered have all related to individual, Darwinian fitness, and traits that affect only their bearer's fitness. As we saw in chapter 1, however, many behaviors and traits exist that also affect the fitness of conspecifics. In the next chapter we shall see how the Price equation can also be applied to expose the logic of social evolution, providing us with the means to explain the evolution of behaviors such as altruism and spite.

CHAPTER FOUR

Inclusive Fitness and Hamilton's Rule

4.1 INCLUSIVE FITNESS EXTENDS CLASSICAL DARWINIAN FITNESS

In chapter 1 we were introduced to the classical notion of fitness involving direct personal reproduction by individuals, conceived by Darwin and Wallace and formalized by Fisher, Haldane, and Wright. In chapter 1 we also saw how some social behaviors pose a problem for the classical view of fitness, and briefly saw how W. D. Hamilton extended it into inclusive fitness, and how he proposed a simple rule that predicts when a social trait experiences positive selection. Here we show, using the Price equation, how inclusive fitness can be mathematically formalized in the simplest case, and how that formalization can be used to derive Hamilton's rule in its simplest form, as applied to unconditional behaviors having additive effects on fitness.

Before we proceed to our first derivation of Hamilton's rule, it is worth emphasizing that Hamilton's rule is one way of applying inclusive fitness theory; this point is worth making since some authors mistakenly identify Hamilton's rule as being synonymous with inclusive fitness theory, which it definitely is not [Marshall, 2011a]. As discussed in chapter 10, various biological phenomena, such as sex allocation and working policing within eusocial insect colonies, have been analyzed by considering what strategies maximize individuals' inclusive fitness, and how observed social behaviors should correlate with quantities such as relatedness;[1] this approach need not involve an explicit formulation of Hamilton's rule.

Let us consider the costs of behaviors to individuals, and the benefits to others, as c and b, respectively; this notation was used for the additive components of fitness in the social games described in chapter 2. If we denote the *relatedness* between individuals, which we will define in the course of this chapter, as r, then a general form of Hamilton's rule that is applicable to additive and unconditional interactions between individuals is

$$rb - c > 0. \tag{4.1}$$

Hamilton's rule in the form of inequality (4.1) states when an unconditional social behavior, that acts additively on fitnesses,[2] experiences positive selection. This form can be interpreted in terms of the direct fitness effects of a behavior $(-c)$, and the indirect fitness effects of that behavior (rb), and is thus simply the statement that the inclusive fitness change arising from a behavior, defined as the sum of its direct and indirect fitness effects (equation (1.2)), is positive. Since b is the effect of a behavior on the fitness of others, r can be interpreted as the "value" that the behaving individual should place on the reproduction of those others, relative to its own [Hamilton, 1972]. Note that although these fitness effects are described in terms of the payoffs of the donation game (table 2.1), in which case b and c are both positive and the social behavior is altruistic, their sign could be changed to describe any other kind of social interaction, such as spite or selfishness, and condition (4.1) would still correctly predict the response to selection.

Inequality (4.1) is all well and good, but how are we to know that it is the *correct* summary of the direction of selection acting on inclusive fitness? Furthermore, can we arrive at more rigorous descriptions of what its variables are, particularly the relatedness r? To answer both these questions we will deploy the Price equation introduced in the previous chapter, and start by considering how to treat costs c and benefits b.

4.2 FITNESS EFFECTS AS REGRESSION ON GENES

We begin our derivation of Hamilton's rule [Hamilton, 1964a] by introducing some notation for the effects of behaviors on fitnesses of individuals that interact socially, to make explicit precisely how genes (and later phenotypes) affect fitness, and also to give a general form of Hamilton's rule that will apply to any (unconditional, additive) behavior regardless of its details. Hamilton noted, in introducing his theory, that the effects of social behaviors are variates, and need to be treated as such [Hamilton, 1964a]. David Queller suggested the approach of treating fitness effects in social evolution models as linear regression coefficients [Queller, 1992b]; hence we treat fitness effects as the simple[3] regression coefficients arising from the best linear prediction of a focal individual's fitness, as a function of their own *genetic value*,[4] and as a function of their social partners' genetic values. However, we are in general dealing with social interactions between genetic relatives, in which case individuals' and partners' genetic values will be correlated; despite the fact that an individual's genetic value may provide information about social partners' genetic values, in dealing with fitness effects we will treat these genetic values *as if* they were independent. In more complicated social scenarios we will see that the simple regression coefficient approach can break down, requiring alternative approaches including partial regression of fitness effects. For now however, we denote the regression of random variable X on random variable Y, assuming that Y is independent of Z, as $\beta_{XY \perp Z}$, where $\perp$ denotes statistical independence. So, we write an expression for the extended social fitness of an individual in terms of their genetic value G and those of their partners G' as

$$W = \overbrace{G\beta_{WG \perp G'}}^{\text{fitness change due to self}} + \underbrace{G'\beta_{WG' \perp G}}_{\text{fitness change due to others}} . \tag{4.2}$$

Equation (4.2) is a very general formulation of social fitness, in which fitness is decomposed into fitness due to the individual's behavior (the first term), and fitness due to the behavior of the individual's social partners (the second term). There are two main things to note about equation (4.2). First, its description of fitness is simplified, since the full version should include the nonsocial baseline component of fitness, and the intercepts and residuals of the linear regression models.[5] However, since we know that we will be using equation (4.2) in the Price equation, assuming that the residuals of the regressions are uncorrelated with the genetic value of the individual whose fitness we are calculating, the covariance terms involving baseline fitness, the intercepts, and the residuals will disappear in the analysis.[6] Second, equation (4.2) describes the fitness of a single individual in terms of the effects of their genetic values on their own fitness, plus the effects of their social partners on the same individual's fitness. However, inclusive fitness is defined as the sum of the effect on own fitness, and the effect on others' fitness weighted by relatedness. Equation (4.2) is what is known as *neighbor-modulated fitness*,[7] and was first used by [Hamilton, 1964a]. As we shall see, neighbor-modulated fitness is equivalent to inclusive fitness in that we can use it to derive a version of Hamilton's rule that is interpretable in inclusive fitness terms.

Our notation for fitness effects may seem unnecessarily complicated, however it has a number of advantages. First, it gives us a simple formulation of inclusive fitness and Hamilton's rule that can be applied without change to any additive social interaction. We illustrate this by applying it to additive donation games and public goods games below. Second, our notation makes explicit that in this version of Hamilton's rule we are measuring fitness effects as those arising from possession of genes, rather than expression of a phenotype. David Queller has pointed out that much of the confusion over the correctness of Hamilton's rule has arisen from not properly distinguishing between fitness effects expressed in terms of genes and those expressed in terms of phenotypes [Queller, 1992a]; our notation makes this distinction explicit. Finally, our notation also enables us to distinguish between considering additive fitness effects, and averaging over nonadditive fitness effects as we shall do in the next chapter to derive a fully general version of Hamilton's rule.

As promised, let us now illustrate the simple regression approach to fitness effects by applying it to a couple of additive social interactions from chapter 2: the additive donation game, and the additive public goods game. Taking the additive donation game first, let us assume that individuals interact once during their lifetime, in pairs, and each receive fitness payoffs according to their own behavior and their social partner's behavior, as described in table 2.1. If behaviors depend linearly and exclusively on genetic values, then an individual's fitness can thus be written as

$$W = -Gc + G'b, \tag{4.3}$$

where G is the focal individual's genetic value and G' is that of their single social partner. Calculating the simple regression coefficients describing fitness effects of own and partner's genetic value gives $\beta_{WG \perp G'} = -c$ and $\beta_{WG' \perp G} = b$, as illustrated in figure 4.1.[8]

Similarly, let us consider the additive public goods game of chapter 2, in which individuals contribute, at personal cost c, to a public good b that is shared equally among all N members of the group including the donor. Then, assuming donation depends linearly and exclusively on genetic value, we can write the fitness of an individual within a group as

$$W = -G\left(c - \frac{b}{N}\right) + G'\frac{b}{N}, \tag{4.4}$$

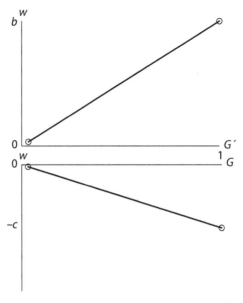

Figure 4.1: Fitness effects can be represented as simple regression coefficients relating fitness to underlying genetic value, making the assumption that genetic values of social partners are uncorrelated. The top plot illustrates the coefficient for own fitness regressed on partner's genetic value ($\beta_{WG'\perp G}$) for the additive donation game (4.3). The bottom plot illustrates the coefficient for own fitness regressed on own genetic value ($\beta_{WG\perp G'}$), for the same social interaction.

where G is the genetic value of the focal individual, and G' is the summed genetic values of all the other members of the group. The regression coefficients describing the fitness effects of the social behavior are then calculated[9] as $\beta_{WG\perp G'} = -c + b/N$ and $\beta_{WG'\perp G} = b/N$.

4.2.1 Correlation versus Causation

Any introductory course on statistical analysis will teach that "correlation is not causation." Probabilistic models of evolution are no different. In constructing a model and then expressing fitness effects in terms of regression coefficients we are making an assumption that these fitness effects capture causal interactions, rather than mere correlations. This can be valid if we know the structure of the underlying model [Okasha, 2006]. It is, of course, perfectly possible to construct models where fitness changes as predicted by aspects of social partners are purely correlative. Benjamin Allen, Martin Nowak, and E. O. Wilson do precisely this and then, in noting the capacity of simple regression-based inclusive fitness models to provide incorrect explanations for certain social scenarios, conclude that inclusive fitness theory itself is at fault [Allen et al., 2013]. Three main points deserve to be made in response to this claim. First, this criticism is not peculiar to inclusive fitness theory, but applies to any regression-based approach to studying evolution, social or otherwise. Second, the approach to analyzing the evolution of correlated asocial characters, which inspired regression approaches to inclusive fitness theory, provides methods for disentangling direct from indirect results of selection on correlated phenotypes [Lande and Arnold, 1983]. Third, as discussed in chapter 10, inclusive fitness is a concept that is not synonymous with the regression approach, which is a tool; although the regression approach is used fruitfully in this book to analyze appropriately structured scenarios, even if we do find particular applications of it to particular scenarios that give misleading evolutionary explanations, these are failures of a mathematical tool, rather than the underlying concept.

4.3 DERIVING HAMILTON'S RULE IN THE SIMPLEST CASE

In the preceding section we defined what we mean by "costs" and "benefits" of social behavior in simple additive scenarios where genetic value perfectly predicts behavior. We are now in a position to derive Hamilton's rule for such scenarios, including the crucially important relatedness parameter r.

4.3.1 Positive Inclusive Fitness

Our simplest version of something like Hamilton's rule would be to ask under what circumstances inclusive fitness for a behavior is positive. To answer this question we need do no more than substitute our equation describing individual, neighbor-modulated fitness into the Price equation and, assuming faithful transmission of genetic values on average, ask when genetic value positively covaries with reproduction. However, this approach is not particularly useful, since it does not tell us why a behavior experienced positive selection, that is, whether it was due to direct fitness effects or due to indirect effects on relatives. This approach also relies on an assumption that neighbor-modulated fitness is the same as inclusive fitness, in terms of the evolutionary predictions that it results in. To resolve these two issues we shall now derive Hamilton's rule, which separates out direct and indirect fitness effects, in a way that enables us to see that neighbor-modulated and inclusive fitness formulations generate equivalent predictions.

4.3.2 Hamilton's Rule

To derive Hamilton's rule we take our general description of individual, neighbor-modulated fitness (4.2), substitute it into the Price equation (3.3), and ask when the social behavior of interest will experience positive selection (i.e., when $\text{Cov}(G, W) > 0$). This gives us[10]

$$\text{Var}(G)\beta_{WG \perp G'} + \text{Cov}(G, G')\beta_{WG' \perp G} > 0. \tag{4.5}$$

Now we note that the covariance between an individual's genetic value and those of their social partners is symmetric;[11] in other words we could be describing the association between genetic value of individuals, and those of their social partners that affect their fitness (the neighbor-modulated fitness approach), or we could be describing the association between genetic value of individuals, and the genetic value of social partners whose fitness they affect (the inclusive fitness approach). Under the latter approach we care about the effect of the focal individual's behavior on the fitness of social partners, which is the same as the effect of social partners' behavior on the fitness of the focal individual, since the social interaction payoffs we are considering are all symmetric (see chapter 2), and all individuals are assumed to be able to express, as well as experience, social behavior.[12] So, $\beta_{WG' \perp G} = \beta_{W'G \perp G'}$, where W' is the fitness of social partners. Using this fact to rewrite condition (4.5), then dividing through by the variance in G, we get

$$\overbrace{\beta_{WG \perp G'}}^{\text{"cost"}} + \underbrace{\beta_{G'G}}_{\text{relatedness}} \overbrace{\beta_{W'G \perp G'}}^{\text{"benefit"}} > 0, \tag{4.6}$$

which is the simplest form of Hamilton's rule in its inclusive fitness form, applicable to behaviors that depend solely on individuals' genetic values, and that result in additive fitness effects. To relate condition (4.6) back to its first presentation as in condition (4.1), as already described we have the effect on own fitness $-c = \beta_{WG \perp G'}$ and the effect on social partners' fitness $b = \beta_{W'G \perp G'}$. Genetic relatedness r is properly considered as a regression coefficient of social partners' genetic value as predicted by the genetic value of individuals; although Hamilton initially thought that Wright's coefficient of relationship [Wright, 1922], a correlation coefficient, would approximate the correct measure of relatedness [Hamilton, 1964a], he subsequently and correctly identified relatedness as a regression coefficient [Hamilton, 1970, Hamilton, 1972].[13] The approach presented here to deriving fitness effects and relatedness and to switching between neighbor-modulated and inclusive fitness formulations, all using regression coefficients and the Price equation, follows that of David Queller [Queller, 1985, Queller, 1992b].

4.3.3   Estimating Relatedness from Pedigrees and from Genetic Data

Hamilton initially presented relatedness in terms of kinship within family groups, providing a very simple method of estimating r under certain assumptions. He described r as it would be estimated on the basis of a pedigree analysis in a family tree, as "one-half for sibs, one-quarter for half-sibs, one-eighth for cousins," and so on [Hamilton, 1964a]. Hamilton calculated r in terms of Wright's *f-statistics* or, equivalently for infinite population models, in terms of probabilities of alleles within a focal individual, and contained in social partners, being *identical by descent*.[14] For this measure of r to correspond with the degrees of relatedness arising from a simple pedigree analysis, a number of assumptions need to be satisfied including, as Hamilton noted, that selection on the trait in question is vanishingly weak.[15] The use of pedigree analysis in the calculation of relatedness led John Maynard Smith to label inclusive fitness theory as *kin selection* theory, in contrast to *group selection* theory [Maynard Smith, 1964]. Subsequent to Hamilton's original papers, a number of alternative formulations of the coefficient of relatedness were developed. Richard Michod and Hamilton were able to show, again under the assumption of weak selection, that these coefficients were all equivalent [Michod and Hamilton, 1980]. These different versions of relatedness, and their subsequent reconciliation, were all derived using population genetical approaches, without the Price equation.[16] The regression form of relatedness as derived in inequality (4.6), however, made no assumptions of weak selection. Hamilton anticipated this development in the original presentation of his theory, writing,

> Strictly, a more complicated metric of relationship taking into account the parameters of selection is necessary for a locus undergoing selection, but the following account based on use of the above coefficients must give a good approximation to the truth when selection is slow and may be hoped to give some guidance even when it is not. [Hamilton, 1964a]

The version of r derived in condition (4.6) is the generalization that Hamilton pointed out would be needed to deal with strong selection, a point we shall return to below.

As the early work of Hamilton and others showed, relatedness can be reasonably well estimated in a simple way from pedigree analysis when selection is sufficiently weak. Therefore for an experimental biologist wanting to test inclusive fitness theory in a real biological system, one approach to measuring relatedness would be to infer it directly from pedigree analysis. However, if the behavior under investigation were under strong selection, the relatedness estimates thus inferred would typically be inaccurate. Since social behaviors might be expected normally to be under strong selection, how is relatedness to be reliably estimated? David Queller and Keith Goodnight observed further difficulties in using pedigree analysis to infer relatedness, such as difficulties in inferring paternity in many species, and even inferring maternity in some [Queller and Goodnight, 1989]. Queller and Goodnight, generalizing earlier approaches, developed techniques for estimating relatedness directly from genetic data;[17] at their heart these techniques use versions of the regression coefficient form of relatedness given in inequality (4.6), although they can also be interpreted in terms of identity-by-descent [Queller and Goodnight, 1989]. Experimental tests of inclusive fitness theory can make use of both genetic and pedigree analysis approaches to estimating relatedness in populations (e.g., in long-tailed tits [Nam et al., 2010]; see figure 10.2).

4.4 PERCEIVED LIMITATIONS OF INCLUSIVE FITNESS THEORY

Over the decades since its publication, inclusive fitness theory has been the subject of much criticism. A number of apparent criticisms were made in the early decades of the theory. In the few years heading up to the theory's 50th anniversary, some of these criticisms resurfaced, and were joined by entirely new ones. The purpose of this book is to celebrate the generality of inclusive fitness theory, and part of this entails defending it from claims of its limited scope. Rather than revisit old controversies at length, here we will briefly discuss the most pertinent criticisms and their resolution. A fuller treatment of the resurgent criticisms of inclusive fitness theory, which reignited in the early twenty-first century, is given in [Bourke, 2011b, Gardner et al., 2011, Marshall, 2011a, Rousset and Lion, 2011, Birch, 2013].

In brief, the criticisms of inclusive fitness theory that have been made most frequently, or most forcefully, are that inclusive fitness theory is a theory applicable only to behaviors under weak selection, which have additive fitness effects and occur only between pairs of individuals rather than within groups. Because of these limitations, it is claimed that multilevel selection theory is more general, since it is not subject to any of these restrictions, and is thus a distinct and superior approach to analyzing social evolution. These claims are not new, and variants of them have been promulgated over the five decades since the publication of Hamilton's papers, but in most researchers' eyes they came to a head with the publication by *Nature* in 2010 of a paper by Martin Nowak, Corina Tarnita, and Edward O. Wilson [Nowak et al., 2010]. To the preexisting claims for the limitations of inclusive fitness theory and the superiority (and nonequivalence) of group selection, Nowak, Tarnita, and Wilson added the further charge that inclusive fitness was not an extension to classical (Darwin–Fisher) fitness, but rather a confusing reconceptualization of classical fitness.[18] In the following we will address these claims.

4.4.1 Hamilton's Rule and Multilevel Selection Theory Give Equivalent Predictions

The claim that inclusive fitness theory and multilevel selection theory do not make the same evolutionary predictions is based on the misconception that inclusive fitness theory is not fully general, in that it assumes weak selection and additive interactions (points we shall consider below), and applies only to interactions between pairs of individuals (e.g., [van Veelen, 2009, Nowak et al., 2010, Traulsen, 2010]).

   To show that inclusive fitness and multilevel selection models of the same behavior do indeed make the same evolutionary predictions, in terms of the direction of selection on a behavior,[19] we will again make use of the Price equation. We used the Price equation above to derive our inclusive fitness version of Hamilton's rule for additive behaviors (condition (4.6)). Note that this rule can describe the evolution of social behavior within groups, since it was applied to the additive public goods game in section 4.2, hence the claim that inclusive fitness applies only to pairwise interactions is clearly incorrect. Having used the Price equation to derive Hamilton's rule, we use the same starting point to derive a multilevel selection rule. So, we start with the rule $\mathrm{Cov}(G, W) > 0$, where W is neighbor-modulated fitness. We can then rewrite this according to a mathematical rule, the *law of total covariance*,[20] and rearrange to give

$$\overbrace{\mathrm{Cov}(\mathrm{E}(G|N), \mathrm{E}(W|N))}^{\text{between-group selection}} > \overbrace{-\mathrm{E}(\mathrm{Cov}(G, W|N))}^{\text{within-group selection}}, \qquad (4.7)$$

where N is the group size as before. This is the multilevel selection rule for the evolution of a social behavior, described in chapter 1 (condition (1.3)). Let us interpret this rule with reference to donation behavior in the public goods game of chapter 2. On the left-hand side we have the covariance (weighted by group size) between expected genetic value within a group, conditioned on group size, and expected fitness within a group, again conditioned on group size; this is therefore the covariance between group-level genetic value and group-level fitness, and captures the process of between-group selection, by which groups with more donators produce more offspring. On the right-hand side we have the expectation (weighted by group size) of the covariance between genetic value and fitness within groups, again conditioned by group size; this captures the process of within-group selection, wherein donators are at a relative disadvantage within their groups because they pay a fitness cost while all group members share the public good equally. Rule (4.7) tells us that when the magnitude of between-group selection exceeds the magnitude of within-group selection, a social behavior will experience positive selection. George Price, and subsequently Hamilton, were the first to apply his equation to formulate multilevel selection in this way, although Price's approach was recursive,[21] whereas this approach is based on a mathematical identity, achieved by conditioning the covariance term of the Price equation [Marshall, 2011a]. Note that we have interpreted social evolution in causal terms using rule (4.7); Samir Okasha has provided a comprehensive and subtle consideration of multilevel selection, including causal interpretations [Okasha, 2006]. With its focus on inclusive fitness, it is beyond the scope of this book to do the topic of multilevel selection justice. However, the above derivation should satisfy the reader that Hamilton's rule and the multilevel selection rule predict the same selective pressure regardless of group structure and, as we shall see below, regardless of the strength of selection and the additivity of fitness effects. This is because both can be derived from the same starting point: the simple Price equation for neighbor-modulated fitness.

4.4.2   Neighbor-Modulated Fitness Is *Not* Classical Fitness

Mention was made above of the claim that inclusive fitness, in its equivalent neighbor-modulated form, is actually the same quantity as the classical fitness envisaged by Darwin and Wallace, and formalized by Fisher [Darwin and Wallace, 1858, Darwin, 1859, Fisher, 1930]. Under this viewpoint, inclusive fitness becomes a conceptually confusing reinterpretation of classical evolutionary theory, rather than a genuine and insightful extension [Nowak et al., 2010].

Hamilton, however, was very clear that inclusive fitness is an extension of classical fitness, writing,

> In [this] equation, the symbol also serves to distinguish this neighbour-modulated kind of fitness from the part of it...which is equivalent to fitness in the classical sense of individual fitness. [Hamilton, 1964a]

The classical definitions of fitness typically did not explicitly exclude the social environment, as Hamilton did here and in the quotation in chapter 1, probably because, as he noted,[22] the pioneers of the modern synthesis did not often consider social effects, and typically seemed to consider them unimportant. Fisher, for example, said,

> There will also, no doubt, be indirect effects in cases in which an animal favours or impedes the survival or reproduction of its relatives.... Nevertheless such indirect effects will in very many cases be unimportant compared to the effects of personal reproduction. ([Fisher, 1930]; quoted in [Foster, 2009])

The apparent lack of an explicit exclusion of social environment in the classical theory may be part of the source of this confusion. The other part may lie in experimental practice. Before inclusive fitness theory, an experimental biologist would measure classical fitness by counting surviving offspring, unaware of the need under the classical theory to exclude the social effects of others on their production, and unable to distinguish these effects in any case. After Hamilton's work, an experimental biologist might still measure fitness in the same way, aware that they should be measuring inclusive fitness but, due to the frequent impossibility of distinguishing direct and indirect components of inclusive fitness and accounting for them as Hamilton prescribed, settling instead for measuring total individual offspring production, interpreted as neighbor-modulated fitness. Thus the quantity measured would not have changed with the introduction of inclusive fitness theory, but its interpretation would have done.

4.4.3 No Need for Weak Selection or Additive Interactions

As noted earlier in this section, inclusive fitness theory has also been criticized for being applicable only to behaviors under weak selection [Nowak et al., 2010, Traulsen, 2010], and only to additive fitness interactions [van Veelen, 2009,

Nowak et al., 2010]. The use of pedigree analysis to estimate genetic relatedness, by Hamilton and other early researchers in the field, is partly to blame here, although Hamilton did explicitly recognize the need to generalize the coefficient of relatedness to deal with situations of strong selection, as described above. However, in the derivation of the relatedness coefficient in inequality (4.6), no assumptions about the strength of selection were made; indeed, the Price-equation approach taken works for any strength of selection. Early theoretical analyses also claimed that inclusive fitness theory gave exact evolutionary predictions only when applied to additive fitness interactions between social partners (e.g., [Cavalli-Sforza and Feldman, 1978, Boorman and Levitt, 1980]). One approach to obviating this apparent limitation of inclusive fitness theory was also to assume weak selection, enabling nonadditive fitness effects to be approximated with additive versions (e.g., [Grafen, 1985b]). Weak selection can also be useful in constructing simplified inclusive fitness models that are analytically tractable, using tools from evolutionary game theory for example.[23] However, as we will see in the next chapter, Hamilton's rule can be formulated to deal with nonadditive fitness interactions, without the need for weak selection.

4.5 Summary

In this chapter we have seen how a Price-equation approach enables us to derive a simple but general form of Hamilton's rule, applicable to additive social interactions where behaviors are unconditionally expressed according to underlying genetics. When these assumptions are satisfied, no further assumptions are required in applying Hamilton's rule to a given scenario. In particular, the rule applies even under strong selection, and in groups of two or more individuals. We have discussed some of the historical reasons why weak selection has been invoked in statements of Hamilton's rule, such as when inferring genetic relatedness from pedigree relationships. The Price equation has also been used to derive a version of the group selection rule, applicable in the same circumstances, demonstrating that predictions as to direction of evolutionary change are the same under Hamilton's rule, and under the multilevel selection rule. We have defined fitness effects in terms of simple regressions on underlying genes, and highlighted that we should give these a causal interpretation only when they properly capture causal interactions within the scenario modeled. We have also highlighted that inclusive fitness is a genuinely novel extension of the classical fitness studied by Darwin, Fisher, and others.

CHAPTER FIVE

Nonadditive Interactions and Hamilton's Rule

In chapter 2 we were introduced to a variety of models of social behavior. Some of these, such as the additive donation game of table 2.1, capture additive interactions between individuals in which the cumulative effects on an individual's fitness from their own behavior and those of their social partners are simply the sum of all those effects. However, we also saw how nonadditive games can radically change the nature of the selective pressures exerted on behavior, leading to situations in which both social and nonsocial behaviors are stable within a population, or a mixed-population equilibrium exists with social and nonsocial types both represented. These results all assume that populations are completely unstructured, so that individuals interact with "unrelated" individuals on average. In this chapter we examine what happens in such nonadditive interactions when the interactions take place between relatives, and how Hamilton's rule can be extended in two different ways to accommodate such nonadditivity.

5.1 REPLICATOR DYNAMICS FOR INTERACTIONS BETWEEN RELATIVES

To explore the selective pressures on nonadditive behaviors directed towards relatives we will, as in chapter 2, make use of the replicator dynamics (equation (2.4)). In chapter 2 we applied this to nonadditive interactions within pairs of individuals in unstructured populations (equation (2.2)); in this section we need to extend this application of the replicator dynamics to capture interactions within structured populations, so that on average, interactions within the population occur between relatives. How we are to achieve this is not immediately clear, however, given that in chapter 2 relatedness was derived as a regression coefficient capturing the extent to which a focal individual's genetic value is a predictor of their social partners' genetic values. Earlier game-theoretic approaches to analyzing interactions between relatives introduced an assortment parameter; individuals interact with individuals having the same strategy with probability α, and otherwise interact with a member of the population sampled uniformly at random [Grafen, 1979, Queller, 1984]. It can easily be shown that this assortment parameter is formally equivalent to the genetic regression coefficient of relatedness derived in chapter 2.[1] Thus to analyze nonadditive interactions between pairs of relatives we simply extend the fitness equations (2.8) and (2.6) of chapter 2 to give their counterparts for the nonadditive donation game of table 2.2 played between relatives, as

$$w_I = -c + \alpha(b + d) + (1 - \alpha)f(b + d) \tag{5.1}$$

and

$$w_{II} = (1 - \alpha)fb. \tag{5.2}$$

These equations can then be substituted into the replicator dynamics,[2] enabling us to analyze the selective pressures on nonadditive interactions within pairs of individuals, both for single behaviors and for the interesting case where individuals have distinct behavioral roles, and potentially different behavioral strategies for each role.

5.1.1 Simple Nonadditive Interactions within Pairs

For the nonadditive donation game with interactions between relatives we find that, just as for nonadditive games in unstructured populations, mixed-population equilibria can exist. These equilibria and their stability were first described by Alan Grafen and David Queller [Grafen, 1979, Queller, 1984]. Here we can calculate this equilibrium, and then determine its stability, just as we did in chapter 2. If the deviation from fitness additivity d is positive we find[3] that the equilibrium can exist only when

$$\frac{c - d}{b} < \alpha < \frac{c}{b + d}, \tag{5.3}$$

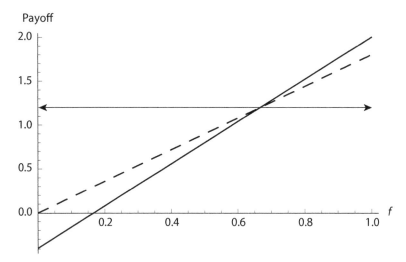

Figure 5.1: Visualization of selection pressures in the positively nonadditive donation game between relatives. Solid and dashed lines show payoffs to action **I** (donating) and action **II** (nondonating) players respectively, as the frequency of type **I** players, f, varies on the x-axis. Higher payoffs correspond to increased success under the replicator dynamics, and the arrows show how f evolves under the replicator dynamics of equation (2.2), away from a mixed-population equilibrium. Parameters are $b = 3, c = 1, d = -1, \alpha = 11/20$.

while if d is negative the equilibrium can exist only when

$$\frac{c}{b+d} < \alpha < \frac{c-d}{b}. \tag{5.4}$$

We next check the stability of the equilibria that can result under positive and negative additivity. We can show that, under certain reasonable conditions,[4] the mixed-population equilibrium is unstable when interactions are positively nonadditive (figure 5.1), and stable when they are negatively nonadditive (figure 5.2) (see also [Queller, 1984]). These outcomes are similar, but not identical, to those presented in chapter 2 for interactions between nonrelatives.

5.1.2 Nonadditive Interactions within Pairs with Roles

The preceding section assumed that behaviors in interactions between relatives were under the same genetic control in both individuals. This need not be the case however and, when it is not, interesting evolutionary outcomes can occur. In this section we will consider a simple description of interactions between relatives where individuals occupy distinct behavioral roles, and where the genetic basis of a behavior depends on which role an individual finds themselves in.

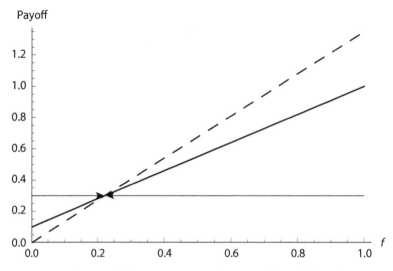

Figure 5.2: Visualization of selection pressures in the negatively nonadditive donation game between relatives. Solid and dashed lines show payoffs to action **I** (donating) and action **II** (nondonating) players respectively, as the frequency of type **I** players, f, varies on the x-axis. Higher payoffs correspond to increased success under the replicator dynamics, and the arrows show how f evolves under the replicator dynamics of equation (2.2), towards a mixed-population equilibrium. Parameters are $b = 3, c = 2, d = 1, \alpha = 4/10$.

Individuals should be expected to express behavior conditional on their state, whether physiological or otherwise, when doing so provides a fitness advantage. David Queller discussed social behaviors that are expressed conditional on state, such as a stronger individual monopolizing reproduction, and forcing weaker relatives to decide whether to stay and act as nonreproductive helpers, or leave and attempt to breed independently [Queller, 1996]. We can readily suppose that in such a situation the social interaction between individuals of different strengths would involve highly asymmetric fitness effects; a weaker individual may never achieve the reproductive success of a stronger individual, for example, no matter how much help they received. Given such asymmetries, one can intuitively see how a social behavior such as donation could become fixed in the population for one behavioral role (such as being the physically weaker member of a pair), whereas nondonation would become fixed in the population for the opposite behavioral role (such as being the physically stronger member). In the following we will see that, even if the fitness effects of behaviors are the same regardless of current behavioral role, one role can become unconditionally associated with donation while the other is unconditionally associated with nondonation.

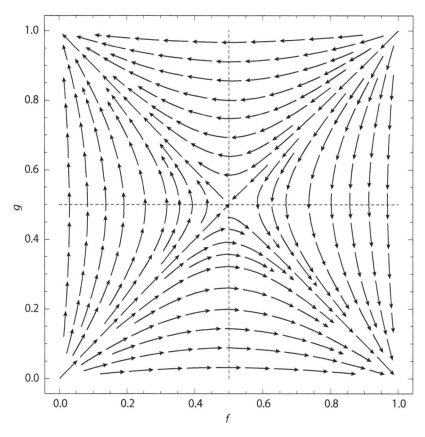

Figure 5.3: Visualization of selection pressures in the negatively nonadditive donation game with roles between relatives. Arrows show the selective pressure experienced under different frequencies of donation (type **I**) behavior in the two behavioral roles. Dashed lines indicate lines of zero selective pressure on each behavioral role, and intersect at the unstable equilibrium. Selection acts to establish donation in one behavioral role, and nondonation in the other, according to initial population frequencies [Marshall, 2009]. Parameters are $b = 5$, $c = 1, d = -1, \alpha = 1/3$.

We consider the model of [Marshall, 2009], in which individuals are randomly assigned to one of two distinct behavioral roles, such that one individual occupies each role.[5] Given payoffs as in the standard nonadditive donation game of table 2.2, and a relatedness coefficient α determining the probability that a social partner is genetically identical to an individual at a behavior-governing locus (see note 1), we can see[6] that selection will act to establish either a population in which individuals always donate in one role, but do not donate in the other (as illustrated in figure 5.3)

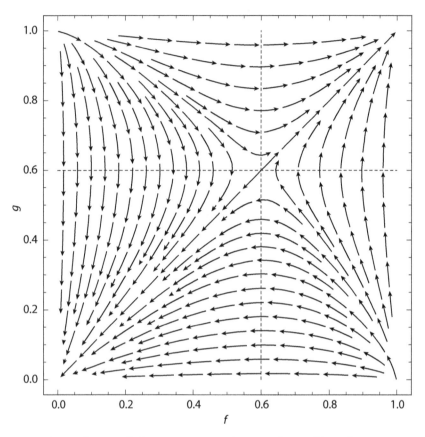

Figure 5.4: Visualization of selection pressures in the positively nonadditive donation game with roles between relatives. Arrows show the selective pressure experienced under different frequencies of donation (type **I**) behavior in the two behavioral roles. Dashed lines indicate lines of zero selective pressure on each behavioral role, and intersect at the unstable equilibrium. Selection acts to establish either donation in both roles, or donation in neither, according to initial population frequencies [Marshall, 2009]. Parameters are $b = 5$, $c = 2$, $d = 1$, $\alpha = 1/4$.

when $d < 0$ and

$$\frac{c}{b} < \alpha < \frac{c-d}{b+d}, \tag{5.5}$$

or a population that donates in both roles or donates in neither (as illustrated in figure 5.4) when $d > 0$ and

$$\frac{c-d}{b+d} < \alpha < \frac{c}{b}. \tag{5.6}$$

Thus additivity of fitness effects controls the selective pressures acting on behaviors in distinct roles, in a similar way to the single-behavior case of equations (5.1) and (5.2). Of particular interest is the finding that negative nonadditivity can give rise to "mixed equilibria" where the whole population donates under one behavioral role but not under the other, as shown in figure 5.3; this may have implications for the evolution of conditional cooperation between relatives[7] [Marshall and Rowe, 2003, Marshall, 2009], for example. Positive nonadditivity, on the other hand, can give rise to bistability, in which the population either reaches fixation for donation in both roles, or in neither role, as illustrated in figure 5.4.

The fixation of prosocial behavior in one behavioral context, but asocial behavior in another, is of particular interest. One may assume that in order for such an evolutionary outcome to occur there must be an asymmetry in the fitness benefits available to individuals under the different behavioral roles, yet this is not the case in the model that gives rise to figure 5.3. In fact John Maynard Smith and Geoff Parker previously showed that asymmetric payoff matrices are not required in evolutionary games in order for such asymmetric evolutionary stable strategies to arise; rather a simple cue, such as which behavioral role an individual finds itself in, can be used to determine the appropriate behavior to exhibit. Maynard Smith and Parker labeled such cues "uncorrelated asymmetries" [Maynard Smith and Parker, 1976], and the result illustrated in figure 5.3 is an example of precisely this effect. The original analysis of uncorrelated asymmetries assumed composite strategies that specified behavior in both roles [Maynard Smith and Parker, 1976], whereas the analysis summarized in note 6 considers selective pressures at two loci, one for each behavioral role, independent of each other. Reassuringly, the two approaches give the same stable evolutionary outcomes [Marshall, 2009].[8]

5.2 EXTENDING HAMILTON'S RULE TO DEAL WITH NONADDITIVITY

In deriving Hamilton's rule (condition (4.6)) in chapter 4, we derived fitness effects of social behaviors as linear regression coefficients, under the assumption that the residuals of these regressions were uncorrelated with the genetic value of a focal individual (see chapter 4, note 6). To be specific, for all social behaviors both by an individual and by social partners, we ask when conditions of the form

$$\text{Cov}(G, \varepsilon_{WG}) = 0 \tag{5.7}$$

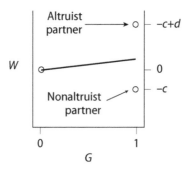

Figure 5.5: Regression of fitness change W on individual genetic value G, demonstrating residuals that correlate with G when interactions are between relatives ($\text{Cov}(G, G') \neq 0$) and interactions are nonadditive ($d \neq 0$). Figure adapted from [Queller, 1992b].

are satisfied, where G is the focal individual's genetic value, and ε_{WG} represents the residuals from some regression of fitness on G. David Queller referred to condition (5.7) as the *separation condition* which predicts when quantitative genetics models give "correct" predictions,[9] and wrote that "the principal cause of nontrivial failures of the inclusive fitness model is nonadditivity of fitness components" [Queller, 1992b], citing his earlier work ([Queller, 1984, Queller, 1985]). That is to say, when one moves from the additive donation game of table 2.1, to the nonadditive donation game of table 2.2, interactions between relatives that lead to correlation between genetic values also lead residuals from simple linear regression models of fitness effects to be correlated with genetic value. This is illustrated in figure 5.5, and we can formally show that this is in fact exactly what happens.[10] As well as identifying the problem, Queller proposed two generalizations of Hamilton's rule to deal with it, as described below.

5.2.1 The Synergistic Coefficient

The first generalization of Hamilton's rule deals with nonadditive fitness effects by packaging them up in a separate term capturing all deviations from additivity of fitness interactions. In its most general form we can write the new description of fitness (excluding baseline asocial fitness) as

$$W = G\beta_{WG \perp G'} + G'\beta_{WG' \perp G} + D, \tag{5.8}$$

where, as in equation (4.2), we omit intercepts and residuals from fitness regression equations, since these will be uncorrelated when plugged into the Price equation (see note 5). In equation (5.8), a new random variable D describes the afore-mentioned deviations from fitness additivity, which may depend on both G and G'.[11] This enables the residuals of the fitness effects regressions to be uncorrelated with fitness once more, since an additional statistical predictor has been added to the description of fitness [Queller, 1992b]. Note however that the more general approach does obscure the dependence of nonadditive effects on partners' genotypes (see note 11).

Calculating when $\text{Cov}(G, W) > 0$ and dividing through by $\text{Var}(G)$, as we did when deriving Hamilton's rule in the previous chapter, gives us a condition for when a social trait experiences positive selection:

$$\underbrace{\beta_{WG \perp G'}}_{\text{effect of own behavior}} + \overbrace{\beta_{G'G} \beta_{WG' \perp G}}^{\text{effect of partners' behavior}} + \underbrace{\beta_{DG}}_{\text{synergistic coefficient}} > 0. \tag{5.9}$$

Since this rule is an extension of Hamilton's rule, containing an additional term, we might refer to it as *Queller's rule* [Marshall, 2011b]. The regression coefficient β_{DG} was described by Queller as the *synergistic coefficient*, because of its similarity with the regression coefficient of relatedness. The synergistic coefficient specifies to what extent, and in what direction, deviation from additivity is predicted by the genetic value of a focal individual. Thus, if β_{DG} is negative then fitness interactions are negatively nonadditive, while if it is positive then fitness interactions are positively nonadditive. We can calculate the synergistic coefficient for the nonadditive donation game[12] of table 2.2, and we could do the same for the nonadditive public goods game of chapter 2.[13]

While the synergistic coefficient solves the problem of correlated residuals, and so allows an inclusive fitness model to be constructed for nonadditive interactions, it introduces a further parameter beyond the original r, b, and c of Hamilton's rule, and may therefore seem like too much of a departure. In particular, are the nonadditive effects captured in D part of a focal individual's direct fitness, or their indirect fitness via their related social partners? An alternative approach is available, however, that seems to stay closer to Hamilton's original thinking.

5.2.2 Fitness Effects as Partial Regression Coefficients

The route to having an inclusive fitness model with only the cost and benefit terms from Hamilton's rule is to move away from simple linear regression models of fitness effects, which require uncorrelated residuals. David Queller's original motivation in identifying the separation condition (5.7) was to show how inclusive fitness models are fundamentally quantitative genetics models, with selective effects decomposable into selection coefficients and heritabilities as in the breeder's equation (3.7). Adding additional complexity to inclusive fitness models, such as introducing the synergistic coefficient to deal with nonadditive fitness interactions [Queller, 1985], is thus a means to maintain the classical quantitative genetics viewpoint on response to selection [Queller, 1992b]. Queller also noted, however, that by moving away from a quantitative genetics approach one can construct inclusive fitness models with only the three parameters of Hamilton's rule, but applicable to any kind of social interaction including nonadditive interactions between relatives[14] [Queller, 1992a]. The approach to this is to use *partial regression coefficients*[15] to describe fitness effects due to own and partners' social behavior. Partial regression coefficients guarantee that linear regression models are constructed with uncorrelated residuals, enabling a fully general version of Hamilton's rule to be derived [Queller, 1992a] as[16]

$$\beta_{WG|G'} + \beta_{G'G}\beta_{WG'|G} > 0, \tag{5.10}$$

where $\beta_{WG|G'}$ is the "average" effect of a focal individual's genetic value on their own fitness (the "cost" c in the original version of Hamilton's rule), and $\beta_{WG'|G}$ is the average effect of the focal individual's social partners on their fitness (the "benefit" b in the original rule).

To illustrate the application of this form of the rule to nonadditive interactions between relatives, we can calculate the partial regression coefficients for the nonadditive donation game (table 2.2) with assortment rate α as [Gardner et al., 2007b]

$$\beta_{WG|G'} = -c + \frac{1}{1+\alpha}(\alpha + (1-\alpha)f)d \tag{5.11}$$

and

$$\beta_{WG'|G} = b + \frac{1}{1+\alpha}(\alpha + (1-\alpha)f)d, \tag{5.12}$$

where f is the usual population frequency of donators.[17] Note that the deviation from fitness additivity d, when it arises, is split between the direct and indirect fitness of an individual [Gardner et al., 2011]. We could make similar calculations of the partial regression coefficients for the nonadditive public goods game, given a probabilistic model for the composition of such groups [Marshall, 2014].[18]

5.3 THE PRICE EQUATION AND LEVELS OF CAUSAL ANALYSIS

In chapter 4 we saw how applying the Price equation to studying inclusive fitness raised an issue of the appropriate level of causal analysis; should one simply ask when inclusive fitness is positive, thereby ignoring questions over whether a particular behavior is altruistic or selfish, for example, or should we "drill down" to look at direct and indirect fitness effects? That question had a straightforward answer; considering simply whether inclusive fitness is positive is less informative than considering direct and indirect fitness effects.

In considering nonadditive interactions in this chapter, however, things have become more nuanced. We have seen how extending the Price-equation approach to deriving Hamilton's rule, to deal with nonadditive interactions between relatives, has given rise to two approaches, both due to David Queller. One of these [Queller, 1985] separates out all deviations from fitness additivity, and calculates how these depend on the genetic value of a focal individual (and those of social partners; see note 11). The other [Queller, 1992a] takes any such nonadditivity and divides it between the direct and the indirect fitness of an individual. Which of these is the better approach then? The answer now depends at least partly on what one is interested in studying. The partial regression approach [Queller, 1992a] may seem more attractive as it has a smaller number of parameters, and most closely matches the form of the original version of Hamilton's rule. However, by no longer having an explicit term for how nonadditivity affects selective pressure, what is underlying evolutionary outcomes may become obscured; as we saw in this chapter, nonadditivity can change evolutionary outcomes when interactants are related. At the same time, to some eyes the question may arise of what fitness effects in the general version of Hamilton's rule really mean if no individual within a population ever experiences precisely those effects, but rather they are calculated by "averaging" across a population. Furthermore, the general version's failure to satisfy the separation condition confounds accurate quantitative genetical predictions of evolutionary change in terms of selection differentials and heritability [Queller, 1992b, Birch and Marshall, 2014]. On the other hand, while the synergistic extension of Hamilton's rule [Queller, 1985] satisfies the separation condition and explicitly highlights the effect of nonadditivity on selection, its interpretation in terms of inclusive fitness may be unclear. In the next chapter we will see how the waters are muddied even further when one realizes that nonadditivity can be recast as a conditional expression of behavior, and vice versa. With the proliferation of approaches to modeling social evolution, all of which make the same predictions but differ in their interpretation, a very important class of questions, to do with evolutionary causes for observed outcomes (e.g., [Frank, 1997, Frank, 1998]), have become a source of controversy in themselves. In chapter 7 we shall see the details of this controversy, as well as a potential resolution. We shall also revisit questions of causation and agency in the closing chapters.

5.4 SUMMARY

In this chapter we have seen the effects of nonadditive social interactions on selective pressures when personally costly interactions take place between genetic relatives. When interactions take place between relatives, and their relatedness is at an intermediate level, then two outcomes are possible. If interactions are positively nonadditive then prosocial individuals must exceed a threshold frequency to take over the population, just as when interactions take place between nonrelatives. However, when interactions are negatively nonadditive, then a stable intermediate frequency of prosocial players can arise. These results should generalize to group sizes larger than two. When individuals occupy distinct behavioral roles and have separate genetically specified strategies for each, then similar results hold for intermediate relatedness, with positive nonadditivity leading to genes for social behavior either being fixed in both roles or neither, while negative nonadditivity results in genes for social behavior becoming fixed in one role, but not the other. Such results correspond to the classical correlated asymmetries of evolutionary game theory.

This chapter has also introduced two extensions to Hamilton's rule to deal with nonadditive interactions. One approach takes deviations from additivity and accounts for them all in a single synergistic coefficient. The other approach applies partial regression to keep a version of Hamilton's rule with only three parameters, in which costs and benefits vary according to the frequency of social individuals in a population. The differences between these two approaches, and others, will be considered in chapter 7.

CHAPTER SIX

Conditional Behaviors and Inclusive Fitness

6.1 IMPLICIT AND EXPLICIT CONDITIONALITY

In the preceding chapters we have, with one exception, considered models in which social behavior is unconditionally expressed. The exception was to consider separate strategies for distinct roles that an individual might find themselves in. David Queller argued convincingly that inclusive fitness analyses need to be done on a per-behavior basis [Queller, 1996], citing as an example the decision over whether to reproduce directly, and whether to aid a reproductive. Queller showed that inclusive fitness predictions are only sensible when one analyzes what an individual should do, *given it finds itself in a particular behavioral role*. Various factors might determine the behavioral role an individual finds itself in, such as having a territory or not, being stronger or weaker, or having been raised as a queen rather than a worker within a eusocial insect colony. Thus environmental factors determine the possible behavioral options, and the possible fitness effects arising, for an individual; note that environment may include the genetic environment,[1] in the case of being physically weaker for example, or the epigenetic environment, in the case of being raised on a diet of royal jelly during development [Chittka and Chittka, 2010]. Although possible fitness effects of different behaviors in these examples may vary with the behavioral role occupied, possibly by a large magnitude, in chapter 5 we also saw that "uncorrelated asymmetries" combined with nonadditive fitness effects can lead selection to establish different behaviors for different roles, even when possible fitness payoffs do not vary with role. We might term such conditionally expressed behaviors as *implicitly conditional*,[2] since their expression depends on

71

the environment they occur in but, given that environment, their expression is unconditional.

Another important class of conditional behaviors requires our attention, however; these are behaviors that are differentially expressed, or effective, conditional on some aspect of the phenotype of potential social partners. Such conditionality has typically been suggested as a route to establish prosocial behavior with minimal or no structuring of interactions. Famous examples include Hamilton's *greenbeard* traits, and Robert Trivers's theory of *reciprocal cooperation*. In contrast to implicitly conditional behaviors, we might term the expression of these behaviors as being *explicitly conditional*. Explicitly conditional traits are interesting because it can be unclear how to interpret them using inclusive fitness theory with the result that, as we shall see in this chapter, they have been interpreted in different ways by different researchers. In some extreme cases this has led to proposals that genetic relatedness is unnecessary for altruism to evolve.

6.1.1 Greenbeards

Hamilton initially discussed genetic relatedness primarily as if generated by interactions within kin groups, and recognized that an altruistic individual could profit, in inclusive fitness terms, if they could discriminate close kin from distant kin, and preferentially target altruism towards the former [Hamilton, 1964a]. Furthermore, from the very beginning Hamilton recognized that genetic associations between interacting individuals could be generated even in completely unstructured populations, writing,

> In situations where relationship is not variable...there still remains a discrimination which, if it could be made could greatly benefit inclusive fitness. This is the discrimination of those individuals which do carry...the behaviour-causing genes from those which do not.... At the simplest we need to postulate something like a supergene affecting (a) some perceptible feature of the organism, (b) the perception of that feature, and (c) the social response consequent upon what was perceived. [Hamilton, 1964a]

Richard Dawkins was subsequently to dub such traits *greenbeards*, with reference to a hypothetical *pleiotropic* trait simultaneously encoding for a conspicuous phenotypic marker, such as a green beard, and the direction of altruism exclusively to other individuals bearing green beards [Dawkins, 1976]. Although initially presented as a thought experiment with no supposed link to biological reality, traits that qualify as greenbeards have now been discovered, such as *FLO1*. This trait in yeast codes for a costly protein causing adherence to other *FLO1*-bearers, with corresponding fitness benefits to all members of the resulting biofilm-like structure [Smukalla et al., 2008]; other examples are reviewed in [Gardner and West, 2010]. As we shall see below, opinion differs over whether greenbeard traits should be classified as examples of altruism, or mutual benefit (see table 2.3).

6.1.2   Reciprocal Cooperation

Not long after Hamilton's first papers on inclusive fitness theory were presented, Robert Trivers suggested that repeated social interactions between individuals, and expression of behavior conditional on the previous behavior of others, might provide a route to the evolution of stable prosocial behavior[3] [Trivers, 1971]. The political scientist Robert Axelrod subsequently popularized formal developments of Trivers's ideas, especially by championing the famous *Tit-for-Tat*[4] strategy of starting off by behaving prosocially towards any new social partner, then acting towards them as they themselves acted in the previous "round" of the interaction [Axelrod, 1984]. Axelrod showed how, given a particular version of the donation game known as the prisoner's dilemma, a cluster of rare Tit-for-Tat players could invade and be evolutionarily stable in an otherwise unstructured population, providing they were sufficiently likely to interact with each other and sufficiently numerous [Axelrod, 1984].[5] Axelrod also argued, with Hamilton, how conditional cooperation might become established within populations having low or no relatedness, from initial situations of unconditional altruism within structured populations (but see chapter 5, note 7) [Axelrod and Hamilton, 1981].

6.2   MODELING CONDITIONAL BEHAVIOR

In order to understand the role that conditionality has in the evolution of social behavior, and ultimately to resolve such issues as whether greenbeard traits are altruistic or mutually beneficial, we need to consider different ways in which we can extend our existing unconditional (or implicitly conditional) models to capture explicit conditionality. Two main strategies suggest themselves: modeling conditional expression of behavior through explicit phenotypes, or modeling conditionality as departures from additivity of fitness effects when individuals interact.

6.2.1   Conditional Expression of Behavior

Perhaps the most natural route to modeling conditional behavior is to separate phenotype (behavior) from underlying genotype (genetic value), thereby allowing the former to be conditionally expressed, while defining fitness effects as those that occur *when a behavior is expressed*. This route was first taken by David Queller [Queller, 1985]. For the additive donation game of table 2.1 we could thus write an individual's neighbor-modulated fitness as

$$W = P\beta_{WP \perp P'} + P'\beta_{WP' \perp P}, \tag{6.1}$$

where $\beta_{WP \perp P'}$ and $\beta_{WP' \perp P}$ are respectively the cost $-c$ and benefit b arising from social behavior in the donation game. This approach to describing fitness effects is identical to that used in chapter 4 in the derivation of Hamilton's rule, except that there the regression of fitness was on genetic value, whereas here it is on phenotypic values P and P'; hence the regression coefficients now describe the costs and benefits of behavior *when expressed*. Any rule could define the relationship between an individual's genetic value G and its expressed behavioral phenotype P, hence equation (6.1) could be used to describe greenbeard traits or reciprocal altruism, as described below.

Now, we know from chapter 2 that to determine the direction and strength of selection on a genetic trait using the Price equation, we need to calculate $\text{Cov}(G, W)$ for that trait. Doing this for the general additive fitness model of equation (6.1), dividing through by $\text{Cov}(G, P)$, and asking when the resulting inequality is positive, indicating that the trait receives positive selection, gives us

$$\beta_{WP \perp P'} + \frac{\text{Cov}(G, P')}{\text{Cov}(G, P)}\beta_{WP' \perp P} > 0. \tag{6.2}$$

Inequality (6.2) looks something like the standard version of Hamilton's rule (4.1), but as well as describing costs and benefits in terms of expressed behavior rather than underlying genetic value, the familiar genetic regression coefficient definition of relatedness does not appear; rather a ratio involving the covariance between an individual's genetic value and the expressed phenotypes of themselves and their social partners takes its place.

David Queller first suggested that a condition similar to (6.2) could be applied to investigate the evolution of reciprocal cooperation. Jeffrey Fletcher and Martin Zwick subsequently did precisely this for repeated interactions of fixed length [Fletcher and Zwick, 2006], while [Marshall, 2011c] did the same for interactions of variable length, generalizing classic conditions first presented by Robert Axelrod

[Axelrod, 1984] and Bill Hamilton [Axelrod and Hamilton, 1981] for Tit-for-Tat to invade and remain stable in purely noncooperative populations.[6] Similar analyses can be done even more easily for simple greenbeard traits.[7] We can also extend such analyses to situations where phenotypic markers do not provide unambiguous information about the genetic value of another individual; this is the case where there is no longer a single gene controlling expression both of a conspicuous phenotypic marker and conditional prosocial behavior towards bearers of that marker, but rather when these two traits are controlled by two different genes, that happen to be closely *linked* on the chromosome or otherwise maintained in *linkage disequilibrium*.[8] Considering the potential for phenotypic markers and behavior to be dissociated shows the vulnerability of greenbeard traits to partial suppression or replacement with *falsebeards* [Biernaskie et al., 2011].[9]

6.2.2   Conditionality as Synergy

As mentioned at the beginning of this chapter, explicitly conditional expression of behavior can also be translated into unconditional behaviors but with nonadditive fitness effects. It is quite easy to see intuitively how this can be done: behaviors that are conditionally expressed dependent on phenotypic aspects of an individual's social partners could, instead, be represented as unconditional behaviors whose fitness effects vary nonadditively according to the genetic values of those social partners. This can easily be done for the greenbeard scenario described above,[10] and was done by Jeff Fletcher and Martin Zwick for reciprocal cooperation [Fletcher and Zwick, 2006]. Fletcher and Zwick note that under this viewpoint it appears that instead of a sufficiently large probability of continued interaction making the reciprocally cooperative strategy Tit-for-Tat able to invade in unstructured populations, rather a sufficiently large positively nonadditive payoff benefit when two Tit-for-Tat players interact leads to the same evolutionary outcome. The fact that both the evolutionary success of greenbeards, and the evolutionary success of reciprocal cooperation, can be explained in similar ways, both under a conditional phenotypes viewpoint and a nonadditive fitness effects viewpoint, might suggest that in fact Tit-for-Tat is a type of greenbeard trait itself. This seems reasonable, although the classic greenbeard and Tit-for-Tat should be classified in slightly different ways; the classic greenbeard trait introduced by Hamilton and Dawkins is what Andy Gardner and Stuart West refer to as a *facultative helping greenbeard*,[11] in that it expresses only its prosocial phenotype, and hence pays the fitness cost associated with donation, when it interacts with other greenbearded individuals who necessarily contain the same gene for donation. Tit-for-Tat, on the other hand, donates in the first round of an interaction with a new individual, regardless of their type; if the other individual is not a Tit-for-Tat player they may not donate during that same first round, hence the Tit-for-Tat player is obliged to pay a fitness cost when it begins a new interaction, and is unable to target this initial donation exclusively at other individuals with the Tit-for-Tat "gene." However, if the other player does turn out to be a Tit-for-Tat player then this initial cost is subsequently offset by a nonadditive fitness benefit. Because of this, Tit-for-Tat can be considered as what Gardner and West would term an *obligate greenbeard*, which always pays the cost of donation, with the benefits of this donation being fully enjoyed only by other Tit-for-Tat playing individuals. Jay Biernaskie and colleagues make a similar observation, but also note practical problems with reciprocal cooperation evolving for greenbeard-like reasons in realistic biological scenarios [Biernaskie et al., 2011].

It is worth noting in passing that, in introducing his conditional phenotype model of social evolution, at the same time David Queller also introduced non-additive fitness effects where those effects are expressed in terms of phenotypes [Queller, 1985]. Intuitively, however, we can see how these should not both be necessary to describe any scenario, since nonadditivity can be moved between the genotype–phenotype relationship and the phenotype–fitness relationship in order to ensure that at least one is additive;[12] this was illustrated above when greenbeard traits and reciprocally cooperative traits were analyzed both in terms of conditionally expressed behaviors, and in terms of nonadditive fitness effects. Thus, while a real biological scenario may involve both conditional behavior and nonadditive fitness effects, we do not need all of this complexity to analyze it.

6.3 CLAIMS THAT ASSORTMENT IS MORE FUNDAMENTAL THAN RELATEDNESS

Since some behaviors are conditionally expressed, extensions of Hamilton's rule that deal with them explicitly appear to be more general than the "standard" purely genetic versions [Queller, 1985, Fletcher and Zwick, 2006]. However, the conditional-phenotype extension of Hamilton's rule (condition (6.2)) does not appear to include genetic correlation in its formulation, replacing it with a form of genotype–phenotype assortment; what appears to predict selective pressure in this rule is how possession of a trait for social behavior covaries with the social behavior of others. Since the evolution, even of greenbeard traits, which appear to be classified as altruistic according to the standard criteria (table 2.3), can be predicted by this seemingly nongenetic rule, this has motivated some like Jeff Fletcher, Martin Zwick, and Michael Doebeli to claim that genetic relatedness is not necessary for the evolution of altruism, in stark contrast to the original insights of Hamilton [Fletcher et al., 2006, Fletcher and Doebeli, 2009]. Yet we may be suspicious of such claims; previous theorists claimed that Hamilton's rule was incorrect for nonadditive fitness effects but, as we saw in the previous chapter, a general version of Hamilton's rule with fitness costs and benefits derived correctly can handle such nonadditivity without problem. In the next chapter we will consider the substance of claims that altruism can evolve without genetic relatedness, that genotype–phenotype assortment is a more general quantity than genetic relatedness and, as an aside, whether greenbeard traits should be classified as altruistic (something that inclusive fitness theorists themselves disagree on).

6.4   SUMMARY

This chapter has considered how to model the many social behaviors that may be conditionally expressed. A distinction has been drawn between implicitly conditional traits, and explicitly conditional traits that take account of aspects of social partners. Two classic examples were presented: greenbeard traits and reciprocal cooperation. We saw an extension of Hamilton's rule to deal with explicitly conditional traits, that features a measure of phenotypic assortment that appears not to be the classic genetic relatedness of Hamilton's rule. Arguments that this means that relatedness is not necessary in order for altruism to evolve were briefly presented, and these will be addressed in the next chapter. We also saw that conditional expression of phenotype can be recast, mathematically, as nonadditive-but-unconditional interaction.

CHAPTER SEVEN

Variants of Hamilton's Rule and Evolutionary Explanations

7.1 VARIANTS OF HAMILTON'S RULE

In the preceding chapters we saw how, in response to the apparent challenges posed by nonadditive fitness interactions and conditional behaviors, Hamilton's rule has been generalized to give a number of variants. In this chapter, building on the earlier chapters, we will explicitly separate out four alternative ways of analyzing the evolution of social traits, and show how when applied to the same social behavior they can appear to give different causal explanations for its evolution. This will explain differing perspectives on the evolutionary explanation for positive selection on greenbeard traits, for example, but will also explain a more fundamental point: how, as described in the previous chapter, some authors have come to the conclusion that personally costly altruism does not need genetic relatedness in order to evolve [Fletcher et al., 2006, Fletcher and Doebeli, 2009]. This second point is crucially important to address because, if true, it would indicate that inclusive fitness theory is not the general explanation of social evolution that we suppose it to be. Having identified how different routes to analyzing social evolution, via variants of Hamilton's rule, can lead to potentially vastly different conclusions, we will see how the different approaches can be reconciled. Crucially, we will see how the nongenetic explanation of the evolution of altruism can actually be recast in a version with genetic relatedness. This will then leave us with the more philosophical problem of which approach to analyzing social evolution is closer to revealing its

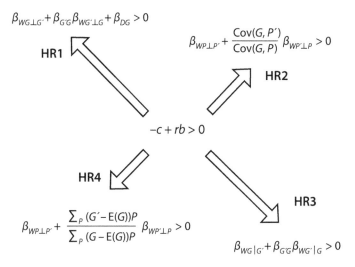

Figure 7.1: Four variants of Hamilton's rule able to accommodate nonadditive and conditional social interactions. HR1 is based on David Queller's synergistic extension to Hamilton's rule [Queller, 1985] (see condition (5.9)). HR2 is based on the same paper by Queller, in which he also introduced conditionally expressed phenotypes in the construction of inclusive fitness models (see condition (6.2)). As discussed in chapter 6, either the synergism term *or* conditionally expressed phenotypes are sufficient to model traits that are both conditional *and* nonadditive. HR3 is David Queller's generalization of Hamilton's rule using partial regressions of fitness on underlying genetic values [Queller, 1992a] (see condition (5.10)). HR4 is Alan Grafen's "geometric view of relatedness" [Grafen, 1985a]. While all these variants of Hamilton's rule may be applied to a given social trait, they may differ in the evolutionary explanation they suggest for it.

"true" underlying causes, and the more practical problem of which is easier to apply to real biological situations.

In the following we will see four different extensions of Hamilton's rule, each of which is able to deal with nonadditive fitness effects and conditional expression of behavior. For ease of reference we shall henceforth refer to these rules, which are presented in figure 7.1, as HR1 through HR4. The first three of these are simply the rules we have already seen in earlier chapters. We have not yet seen the fourth rule, however, and it will turn out to provide a very useful viewpoint. The defining features of the rules are presented in table 7.1.

7.1.1 HR1: Synergism

Our first rule, HR1 (see figure 7.1), is based on one of the first formal generalizations of Hamilton's rule to have appeared in the literature. The rule extends Hamilton's rule with a synergistic coefficient capturing the deviation from additivity of fitness interactions, and was first presented in chapter 5 (condition (5.9)) based on the work of David Queller [Queller, 1985].[1] Under this approach to analyzing social evolution, any deviations from additivity of fitness interactions, conditionality in expression of behavior,[2] or a combination of the two, are captured in a single regression coefficient; this regression coefficient captures the discrepancy between the linear prediction of individual, neighbor-modulated fitness in terms of an individual's genetic value and those of their social partners, and their realized individual fitness.

7.1.2   HR2: Conditional Expression of Phenotype

The second variant of Hamilton's rule, HR2 (see figure 7.1), is also based on the early work of David Queller [Queller, 1985],[3] and was presented in chapter 6 (condition (6.2)). This variant explicitly deals with conditionality in the expression of behavior, which may also account for nonadditive fitness interactions as described in chapter 6, note 12. Variants HR1 and HR2 thus provide complementary approaches to analyzing any combination of conditionally expressed behaviors and nonadditive fitness interactions. The defining feature of HR2 is that, unlike all the other variants presented here, the familiar genetic regression coefficient of relatedness does not appear; instead it is replaced by a covariance ratio describing assortment between genetic values of individuals, and the expressed social behaviors of individuals they interact with.

7.1.3   HR3: Fitness as Partial Regression

We introduced the third variant of Hamilton's rule, HR3 (see figure 7.1), in chapter 5 (condition (5.10)). This version of Hamilton's rule is also due to David Queller [Queller, 1992a] and, by deriving fitness effects in terms of partial regressions of an individual's fitness on their own genetic value and those of their social partners, is able to capture any conditional or nonadditive behavior in the standard three parameters of Hamilton's rule; one of these is the usual genetic regression coefficient of relatedness, unlike HR2.

7.1.4   HR4: The Geometric View of Relatedness

The fourth variant of Hamilton's rule, HR4 (see figure 7.1), is due to Alan Grafen and is usually referred to as the *geometric view of relatedness* [Grafen, 1985a]. Grafen showed how a general version of Hamilton's rule can be derived with a particular form of genetic relatedness which involves summing, over all the instances in which social behavior is expressed, the difference between the genetic value of the donor and the population mean genetic value, and the difference between the genetic value of the recipient and the population mean genetic value. The ratio of these two sums is then taken to be the relatedness required for Hamilton's rule, although it does not at first sight look like the genetic regression coefficient formulation of relatedness that we have become familiar with in the preceding chapters. Grafen's derivation of Hamilton's rule is thus

$$\beta_{WP \perp P'} + \frac{\sum_P (G' - \mathrm{E}(G))P}{\sum_P (G - \mathrm{E}(G))P} \beta_{WP' \perp P} > 0. \tag{7.1}$$

The version of Grafen's inequality (HR4) above is slightly more general than the version he presented, since it allows any level of social behavior from none (no expressed phenotype: $P = 0$), upwards ($P > 0$).[4] The original presentation [Grafen, 1985a] considered only two possible behaviors: donation ($P = 1$) or nondonation ($P = 0$). This simple scenario provides the easiest route into understanding the geometric view, since under this scenario we can see that relatedness is calculated only for scenarios in which the social behavior is expressed by the focal individual ($P = 1$), in which case it is calculated as the average[5] distance between the genetic value of recipients and the population mean genetic value, normalized by the average distance between the genetic value of donators and the population mean genetic value. Thus relatedness can be conceptualized as the relative distance that a recipient's genetic value lies along the line between the population mean genetic value, and the donor's own genetic value [Grafen, 1985a]; this geometric view is illustrated in figure 7.2. As well as providing an intuitive picture of relatedness and why it matters, recall that the geometric view underlies the statistical techniques developed to estimate relatedness from genetic data [Queller and Goodnight, 1989], as discussed in chapter 4. Rather than reproduce Grafen's derivation of HR4, we shall convince ourselves later in the chapter that condition (7.1) is indeed a sound statement of when natural selection favors a social behavior by showing that it predicts the same direction of evolutionary change as other HR variants we have previously derived.

The generalized version of the geometric view of relatedness is applicable to any kind of behavior where there is a linear relationship between the level of behavior expressed and the fitness effects on self or others (and as noted in

Figure 7.2: Under the geometric view in its simplest form, relatedness is defined as the relative position of the genetic value of a potential recipient of social behavior (R), along a line from the population mean genetic value (μ) to the potential donator's own genetic value (A). In the example shown, the potential recipient is halfway from the population mean to the potential actor, hence the relatedness of the actor to the recipient is $1/2$. If the potential recipient R were at the population mean μ, then relatedness would be 0, whereas if the potential recipient were at A, and thus had the same genetic value as the potential actor, relatedness would be 1. Figure redrawn from [Grafen, 1985a].

chapter 6, note 12, any nonlinear relationship between behavior and fitness effects can be recast as a further level of conditionality in the expression of the behavior in the first place). Thus it is applicable to reciprocal cooperation scenarios. The geometric view can be thought of as breaking relatedness down on a case-by-case basis, according to expressed phenotypes; indeed, [Nee, 1989] showed rigorously how one can move from HR2 to arrive at HR1, deriving David Queller's synergistic coefficient for reciprocal cooperation along the way.[6] Nee also showed that when individuals express the same phenotype, geometric relatedness between them is always 1 [Nee, 1989].

7.1.5 Summary of the Variants

The different variants of Hamilton's rule differ in how they treat the three key parameters of the original: "relatedness," "cost," and "benefit" (respectively r, c, and b in figure 7.1). In some variants, fitness effects are treated in terms of expressed phenotype, while in some they are treated in terms of underlying genetic values. In some variants, fitness effects are simple, independent regression coefficients; in others they are derived from partial regressions. In some variants, genetic relatedness features, while in others assortment between genetic value and expressed phenotype is used. These differences are summarized in table 7.1. As one might expect, as a result the different variants of Hamilton's rule appear to give different evolutionary explanations when used to analyze particular social traits, as we shall see later in this chapter. As we shall also see, we can translate between variants that appear, on first sight, to be vastly different. This should be unsurprising given all the variants are derived from the same starting point: the Price equation.

Rule	Relatedness	Cost/Benefit	Notes
HR1	Genetic, regression coefficient	Simple regression on genetic value	Conditionality and nonadditivity captured in "synergistic coefficient"
HR2	Genotype–phenotype covariance	Simple regression on phenotype	Nonadditivity captured in conditionally expressed behavior
HR3	Genetic, regression coefficient	Partial regression on genetic value	—
HR4	Genetic, geometric	Simple regression on phenotype	Relatedness calculated conditional on expressed phenotype

Table 7.1: The defining features of the four variants of Hamilton's rule, in terms of how the relatedness, cost, and benefit parameters of the original are calculated. Simple regressions assume independence of individual's and partners' genetic or phenotypic values, by definition. "Phenotype" refers to the level of social behavior expressed by self and partners.

7.2 GEOMETRIC RELATEDNESS UNDERLIES PHENOTYPIC ASSORTMENT

As discussed above, some authors have concluded that the generality of the genotype–phenotype covariance form of Hamilton's rule (HR2 in figure 7.1) means that altruism, defined as behavior that is personally costly in terms of lifetime fitness (table 2.3), can evolve without genetic relatedness (e.g., [Fletcher et al., 2006, Fletcher and Doebeli, 2009] and, as we shall see in chapter 8, [Frank, 2013]). Are we to interpret this as a deathblow to the generality of inclusive fitness theory? The answer is no, but different arguments as to why this is the correct answer exist. Here we focus on asking about the genetic associations that underlie genotype–phenotype associations. Since phenotypes arise at least in part from genotypes, intuitively it seems that such genetic associations should exist. Satisfyingly, we can use the same covariance identity that Sean Nee made use of in moving from HR2 to HR1 (identity (3.17) in chapter 3, note 5), and apply it to HR2 to rewrite it, very simply, as HR4 (figure 7.1).[7] Thus the genotype–phenotype covariance approach to analyzing social evolution (HR2) is actually a thinly veiled version of Alan Grafen's geometric view of relatedness (HR4). Any conclusions about what conditions are necessary for the evolution of "altruism" that we draw from using HR2 will probably, given the differences between the Hamilton's rule variants noted in table 7.1, need to be reevaluated when we translate to an analysis using HR1 or HR4, or start an entirely different kind of analysis using HR3. In the following section we will do precisely this for greenbeard traits. Before we do so, however, we should deal with one other approach to arguing that altruism can evolve without genetic relatedness.

7.2.1 "Altruism" without Common Genetics

In arguing that genetic relatedness is not necessary for altruism to evolve, Jeffrey Fletcher and Michael Doebeli introduced a simple thought experiment in which two reproductively isolated populations, each with a different DNA-sequence coding for "altruistic" donation towards members of the other population, are paired by a hypothetical experimenter such that donators from one population are always paired with donators from the other, and vice versa [Fletcher and Doebeli, 2009]. Fletcher and Doebeli argued that donation would experience positive selection in such a scenario and that, since donation in each population would be under control by a different DNA sequence, genetic relatedness could not be the reason for this. This is in agreement with a model by Steve Frank, apparently showing that altruism between species (and hence not having a common genetic basis) can evolve [Frank, 1994].

There is more than one possible resolution of the Fletcher and Doebeli thought experiment, and these resolutions have much in common with the analysis of greenbeard traits that we shall undertake in the next section. Andy Gardner and colleagues, for example, analyze Fletcher and Doebeli's thought experiment in inclusive fitness terms and argue that it is not *genotypic similarity* that matters for inclusive fitness theory, but rather genetic relatedness [Gardner et al., 2011]; in other words, it is not important for inclusive fitness theory whether two traits are encoded by identical sequences of DNA, but rather that they encode the same functional trait and that assortment for this trait structures interactions within the population. Gardner et al. implicitly use the geometric view of relatedness (HR4) under this "genetic similarity" viewpoint to conclude that social behavior in Fletcher and Doebeli's scenario is actually explained as altruism between genetic relatives [Gardner et al., 2011]. This argument echoes one made previously by George C. Williams, that it is not the physical nature of genes that matters, but rather their informational content [Williams, 1992]. Moving beyond the artificialities of the Fletcher and Doebeli thought experiment, Gregory Wyatt, Stuart West, and Andy Gardner constructed spatial population genetical models in which the evolution of social behavior could be interpreted either as altruism between species or altruism within species [Wyatt et al., 2013]; Wyatt and colleagues' analysis was unable to discriminate between the two explanations, but they noted that the fact that interspecies social behavior is always explainable as within-species altruism is consistent with Darwin's famous statement that any instance of between-species altruism "would annihilate my theory, for such could not have been produced through natural selection" [Darwin, 1859].

Alternatively one might note that, as discussed below, greenbeards can be vulnerable to modifier genes that suppress donation behavior towards others while leaving intact the phenotypic marker that triggers donation in social partners [Biernaskie et al., 2011]. Chris Quickfall and colleagues adapt this approach to show that under a wide range of conditions in the Fletcher and Doebeli scenario, selection favors precisely such donation-silencing modifiers [Quickfall et al., 2015]. Stable donation between the two populations can arise, however, and may be even easier to select for if the greenbeard mechanism is one that enforces donation by bearers of the phenotypic marker, such as expression of *FLO1* described in chapter 6.

Stable examples of social behavior between populations where there are no common genetics are ubiquitous in nature of course. Interspecific mutualisms abound, for example, in plant–pollinator interactions, in associations between fungi and photosynthetic algae or cyanobacteria within lichens, and so on. Many of these mutualisms clearly provide immediate direct fitness benefits to both parties, such as a hawkmoth receiving a fitness-enhancing drink of nectar from a flower, while the plant producing the flower and nectar achieves pollination in exchange [Quickfall et al., 2015]. Apparently costly donation between species may also experience positive selection when inclusive fitness benefits are directed towards conspecific relatives, such as descendants, but mediated by immediate donation towards a partner species.[8] This is likely to occur within lichens, for example, where benefits to relatives directed via a partner species will be more usual than direct return benefits to a focal, donating, individual. Such scenarios would thus be classified as *altruism involving species* rather than *altruism between species* [Quickfall et al., 2015]. Analyzing direct and indirect fitness benefits in interspecies mutualism is discussed in more detail in chapter 10. In summary, however, these analyses agree with Hamilton's insight when he observed that "no instance of mutualism yet shows either partner giving benefit for zero or negative return"

[Hamilton, 1996].

7.3 EXPLANATIONS FOR GREENBEARDS

We now examine how the different variants of Hamilton's rule might explain, in evolutionary terms, the positive selection that greenbeards experience. We need to take two approaches to analyzing fitness costs and benefits. The first, used in HR1 and HR3 (figure 7.1), is to undertake the analysis in terms of genetic values, and payoffs in the nonadditive donation game (table 2.2). The second is to undertake the analysis in terms of expressed phenotypic values, and payoffs in the additive donation game (table 2.1); this is the approach used in HR2 and HR4 (figure 7.1).

7.3.1 Fitness Effects as Regressions on Genetic Values

In treating greenbeard traits using regressions directly on genetic values we can either, as discussed in the previous chapter, consider a nonadditive set of fitness effects, or instead we can consider conditionally expressed behaviors with additive fitness effects. Under both approaches we will consider unstructured populations, resulting in an apparent relatedness of zero.[9] We illustrate the former approach in analyzing greenbeard evolution using HR1. Under this approach the payoffs in the nonadditive donation game (table 2.2) are $b = 0$, $-c = 0$, and $d > 0$, since greenbeards (type **I** players) experience no fitness increment or decrement unless their social partner is also a greenbeard, in which case they pay an immediate fitness cost for donation but receive a larger fitness benefit in return from their partner's donation.[10]

If instead we move to additive fitness effects, but with conditionally expressed phenotypes, the analysis is slightly different. We illustrate this approach using HR3, allowing the partial regression formulation of cost in Hamilton's rule to capture the conditional expression of behavior. Under this analysis, we find that this cost is in fact negative.[11]

Regardless of whether one analyzes the fitness effects of Hamilton's rule in terms of simple (independent) regressions on genetic values (HR1), or partial regressions on genetic values (HR3), greenbeards are thus classified as traits that spread through direct fitness benefits to their bearers, without the requirement for genetic relatedness. These conclusions are unchanged even if one considers greenbeard scenarios in which the roles of potential donor and potential recipient are separated, and individuals engage in a social interaction only once during their

Rule	Level of analysis	Relatedness	Cost	Benefit	Explanation
HR1	Genetic	0	0	0	Direct fitness benefit due to positive synergistic payoff
HR2	Phenotypic	1	>0	>0	Altruism due to genotype–phenotype assortment
HR3	Genetic	0	<0	>0	Direct fitness benefit
HR4	Genetic	1	>0	>0	Altruism between genetic relatives

Table 7.2: Possible evolutionary explanations of positive selection experienced by greenbeard traits. The four main variants of Hamilton's rule can differ in the evolutionary explanations they provide of selection for greenbeards, due to differences in the level at which they interpret the relatedness, cost, and benefit terms of Hamilton's rule: either genetic or phenotypic. These differing explanations of the same evolutionary scenario illustrate the problem, as described in the main text, of selecting an appropriate variant of Hamilton's rule to apply to a given social trait.

lifetime[12] (e.g., [Biernaskie et al., 2011]). Under such a scenario, when a greenbeard donates, it does so at a cost to its direct fitness that will never be offset by benefit received from another greenbeard; donation is thus personally costly, in direct fitness terms, to that particular greenbeard, although *on average*, greenbeard bearers enjoy a direct fitness advantage under the HR1 and HR3 analyses (table 7.2).

7.3.2 Fitness Effects as Regressions on Phenotypic Values

It may seem counterintuitive to analyze fitness costs and benefits in terms of underlying genetic value, when costs and benefits arise from conditionally expressed behavioral phenotypes. When analyzing conditional greenbeard donation, we can also derive fitness effects in terms of regressions on expressed phenotypes as in HR2 and HR4 (figure 7.1). Under these approaches, the cost in Hamilton's rule associated with bearing a greenbeard trait is always positive,[13] hence greenbeard donation appears to be an altruistic *behavior*. Positive selection for such behavioral altruism is then explained by a relatedness of 1 in Hamilton's rule.[14] The two variants of Hamilton's rule differ in their interpretation of this relatedness, however: HR2 interprets it as genotype–phenotype assortment, whereas HR4 interprets it as traditional genetic relatedness. As already discussed above, in fact HR4 underlies HR2; we will further discuss the importance of this below.

7.4 DIFFERENT VIEWPOINTS ON CONDITIONAL TRAITS

The different viewpoints on social evolution that the variants of Hamilton's rule offer, and the different evolutionary explanations that they give rise to when applied to the classic greenbeard trait, exemplify different traditions within the social evolution literature. Defining fitness effects using partial regressions on genes (HR3) formalizes what is known as a *mutation test*; such a test, which determines whether a trait is advantageous to its bearer in direct fitness terms, works by "mutating" a focal individual's genetic value while holding everything else constant, then determining whether the individual's direct fitness would increase or decrease as a result (e.g., [Fisher, 1941, Nunney, 1985]). This is precisely what a partial regression on an individual's genetic value achieves, of course. This approach is implicitly used by Kevin Foster and colleagues [Foster et al., 2006b] in responding to claims by Fletcher and colleagues [Fletcher et al., 2006] that genetic relatedness is unnecessary for altruism to evolve; Fletcher et al. in particular claim that reciprocal cooperation is a form of altruism that does not require *genetic* relatedness, whereas Foster et al. point out that being a reciprocal cooperator leads to *direct* fitness benefits, as Hamilton himself noted (see chapter 6, note 3). The fact that the partial regression coefficient approach also classifies greenbeard traits as beneficial in direct fitness terms (HR3 in table 7.2) agrees with the intuition of some of those authors that greenbeards evolve due to direct fitness benefits (e.g., [Foster, 2009]).

Other authors, on the other hand, take a different tack in analyzing greenbeard-like scenarios. In addressing Fletcher and Doebeli's suggestion for how altruism between populations might evolve without genetic relatedness, Andy Gardner and colleagues implicitly take a geometric view (HR4) on relatedness, treating both costs and relatedness in Hamilton's rule as those arising when behaviors are expressed [Gardner et al., 2011]. Under this viewpoint, greenbeards experience positive selection, despite being behaviorally altruistic, because relatedness between donor and recipient is always 1 when the donation behavior is expressed (HR4 in table 7.2). This analysis agrees with Hamilton's own intuition when he initially presented the greenbeard thought experiment [Hamilton, 1964a].

Which is the correct approach to take in analyzing explicitly conditional traits: the mutation test or the geometric view? Although one agrees with Hamilton's original insight, this does not settle the matter. Both certainly appear to be valid views,[15] hence a resolution is not attempted here. Recognizing the formal basis for these divergent views seems a good starting point, however.

Finally, one resolution that *is* possible is the claim that genotype–phenotype assortment is more fundamental than genetic relatedness [Fletcher et al., 2006, Fletcher and Doebeli, 2009]. As this chapter has shown, the traditional genetic "geometric view" of relatedness (HR4) is mathematically equivalent to the genotype–phenotype assortment view of relatedness (HR2). Given that genetic relatedness always underlies geno–pheno assortment, and that genes are the indivisible evolutionary units of selection, it would seem appropriate to prefer this geometric view over its genotype-phenotype equivalent.

7.5 SUMMARY

In this chapter we have summarized variants of Hamilton's rule that deal with nonadditive and conditional interactions in apparently different ways. We have seen how these variants appear to give different evolutionary explanations for selection on such traits, exemplified by an analysis of greenbeard traits. An important reason for these differences is that the rules differ in whether they analyze costs and benefits in terms of phenotypes or underlying genes, and differ in how they measure relatedness. However, we have shown the limitations of claims that altruism can evolve without genetic relatedness, by showing how a version of Hamilton's rule applicable to reciprocal cooperation is actually exactly equivalent to a classic genetic measure of relatedness: the geometric view. We have also discussed other proposals for how altruism can occur without common genetics, such as in between-species interactions, and discussed resolutions of these claims. The conclusion is that genetic relatedness remains fundamental in explaining the evolution of altruism, just as inclusive fitness theory predicts.

CHAPTER EIGHT

Heritability, Maximization, and Evolutionary Explanations

8.1 WHAT DRIVES SOCIAL EVOLUTION?

Bill Hamilton conceived inclusive fitness theory as a general theoretical extension of classical, Darwinian and Fisherian fitness, and also as providing a maximization result on a par with Fisher's fundamental theorem of natural selection, described in chapter 3.[1] In the preceding chapters we have seen that inclusive fitness theory is indeed general, and able to deal with subtleties such as nonadditive fitness effects, and conditionally expressed phenotypes. However, we have also seen that selection based on inclusive fitness gives equivalent predictions to other models of apparently different evolutionary processes, such as multilevel selection. How then are we to know whether inclusive fitness really captures the essence of selection on social behavior? Furthermore, is inclusive fitness really maximized by the action of selection, as Hamilton proposed? In this chapter we will examine answers to these questions. We will also consider the problem of classifying observed social behaviors in terms of their underlying evolutionary explanations. For evolutionary biologists such questions are fundamental, and we will review theoretical frameworks that attempt such classifications.

8.2 SELECTION AND HERITABILITY

Using the Price equation we were able, in earlier chapters, to show that Hamilton's rule predicts when a social trait experiences positive selection. In fact, since inclusive fitness is an extension of classical individual fitness, Hamilton's rule also predicts when an asocial trait experiences positive selection. Yet is positive inclusive fitness the causal explanation for selection favoring a social behavior, or is it merely a by-product of some other evolutionary cause? In particular we have seen in previous chapters that a multilevel selection rule also makes the same predictions as Hamilton's rule, and that a neighbor-modulated fitness version of Hamilton's rule does the same. To some, it may therefore seem that selection based on differential group performance, and selection based on neighbor-modulated fitness, constitute equally valid explanations, with positive inclusive fitness arising only as a correlation with these fundamental evolutionary explanations.[2]

How might we objectively and formally assess claims that a particular model of fitness and selection captures the true underlying evolutionary process? In other words, how can we assess when a model provides a causal explanation for observed evolution? Biologists summarize the requirement for evolution by natural selection to occur, according to Darwin's conception, as being the existence of heritable variations in traits, leading to differential reproductive success.[3] A nice simplified[4] view of the conditions for an evolutionary change under natural selection is given by the breeder's equation, previously discussed in chapter 3, which as a word equation states that

$$\text{response to selection} = \text{selection differential} \times \text{heritability.} \qquad (8.1)$$

As discussed in chapter 3, the "selection differential" in the breeder's equation is the covariance between the phenotypic value of a trait, and fecundity. The (narrow-sense) "heritability" was also defined in chapter 3 as the regression of genetic value on expressed phenotype. This suggests a line of attack for considering whether a given description of evolutionary change captures the underlying causes, by focussing attention on the nature of the heritability of a trait. In the following subsections we will do precisely this for inclusive fitness, neighbor-modulated fitness, and group fitness formulations of selection on a social behavior, asking exactly what the heritability measures, and whether this makes sense biologically.

8.2.1   Heritability and Multilevel Selection

We begin by considering whether multilevel selection, which shows that a trait experiences positive selection when the strength of between-group selection exceeds the strength of within-group selection (see inequality (4.7)), constitutes a causal explanation for social evolution. To undertake a quantitative genetics analysis of this statement, using the breeder's equation, we need to decompose the between-group and within-group selection terms into coefficients of selection, and heritability. David Queller showed that, assuming his separation condition is satisfied (see equation (5.7) and discussion in chapter 5), the evolutionary change under multilevel selection can be expressed in quantitative genetics terms as[5]

$$\Delta E(G) \propto \overbrace{\text{Cov}(\overline{W}, \overline{P})}^{\text{between-group selection}} \underbrace{\beta_{\overline{GP}}}_{\text{between-group heritability}} + \overbrace{\text{Cov}(W_\Delta, P_\Delta)}^{\text{within-group selection}} \underbrace{\beta_{G_\Delta P_\Delta}}_{\text{within-group heritability}}. \tag{8.2}$$

This shows that multilevel selection involves heritability of group-level traits, in some sense, as well as heritability of within-group traits. However, it is not immediately clear exactly what these heritabilities mean. In particular, in what sense are group-level (average) traits heritable? In fact, as pointed out by Samir Okasha [Okasha, 2006], these heritabilities are the heritabilities of *family means* and *within-family deviations* used in quantitative genetics [Falconer and Mackay, 1996]. This is interesting, because we know that these heritabilities are actually functions of relatedness and of *individual-level heritability*[6] β_{GP}, strongly suggesting that selection acting differentially on groups is not the most fundamental causal explanation for observed evolutionary change. Indeed, in quantitative genetics, pure selection at the level of the group, known as *family-level selection*, can be practised; this is among the most extreme forms of group selection possible, in which entire groups of individuals are selected for reproduction according to their mean phenotypic value for some trait, yet even in this scenario it is still individual heritability, weighted by a function of within-group relatedness, that underlies and drives evolutionary change [Falconer and Mackay, 1996].[7] Notwithstanding these observations, Samir Okasha has constructed alternative evolutionary scenarios in which path analyses can give incorrect causal explanations under the inclusive fitness *or* multilevel selection viewpoints [Okasha, 2014].

8.2.2 Heritability and Neighbor-Modulated Fitness

Having examined the heritability of phenotypes under the multilevel selection viewpoint, we now turn attention to inclusive and neighbor-modulated fitness formulations. We start with neighbor-modulated fitness which, as discussed earlier in this chapter, has been proposed to explain the evolution of social traits. To constitute an evolutionary explanation, in particular the neighbor-modulated fitness approach must provide a meaningful formulation of heritability. What form, then, does heritability take under the neighbor-modulated fitness viewpoint? To answer this question, we take as our starting point the phenotypic version of Hamilton's rule (condition (6.2), referred to as HR2 in chapter 7). By a simple manipulation,[8] the expected evolutionary change under the neighbor-modulated fitness viewpoint can be rewritten in quantitative genetics terms as [Queller, 1992b]

$$\Delta E(G) \propto \mathrm{Cov}(W, P \perp P')\beta_{GP} + \mathrm{Cov}(W, P' \perp P)\beta_{GP'}. \qquad (8.3)$$

In equation (8.3) the covariances are respectively the effect on the focal individual's fecundity of their own behavior and their partners' behavior, with the effect of each individual's behavior considered independently from that of others. Thus, in quantitative genetics terms, these covariances are the selection differential due to own and partners' behavior [Queller, 1992b]. Things are rather more curious when we consider the regressions of genetic values on phenotypes, which correspond to heritabilities (see chapter 3, note 27); the first heritability makes sense since it is the heritability of the focal individual's behavior, but as David Queller writes, "The second heritability can be viewed as the heritability of [partners'] behaviour through its effects on [the focal individual]" [Queller, 1992b]. Thus it is the phenotypes of social partners that are inherited through an individual's offspring, as well as its own. While this partitioning of heritabilities works technically, in causal evolutionary terms it is somewhat confusing, since something that need not be causally due to a focal individual's particular genetic value, their partners' phenotypes, is considered to be inherited by that individual's offspring. As we shall see in the next section, a similar argument raises problems for neighbor-modulated fitness when one considers what quantity individuals, or natural selection, should maximize. If, however, we make the simple and formally equivalent switch to inclusive fitness theory, do things make any more sense in evolutionary terms?

8.2.3 Heritability and Inclusive Fitness

Moving from the neighbor-modulated to the inclusive fitness viewpoint is very simple when populations are not class structured, since as described in chapter 4 (note 12) we can simply move to accounting for all the effects a focal individual has on others, rather than accounting for all the effects that others have on the focal individual. Our revised, inclusive fitness viewpoint on selection and heritability is now

$$\Delta E(G) \propto \mathrm{Cov}(W, P \perp P')\beta_{GP} + \mathrm{Cov}(W', P \perp P')\beta_{G'P}. \qquad (8.4)$$

Equation (8.4) makes much more sense as a causal interpretation of evolutionary change. This is now split into the effect of an individual's behavior on their own fitness, modulated by the heritability of that behavior via their offspring (the first term in equation (8.4)), and the effect of an individual's behavior on their social partners' fitness, modulated by the heritability of that individual's behavior via the offspring of their social partners. Although the mathematical expression for total evolutionary change is equal to that for neighbor-modulated fitness (8.3), now *all* the fitness effects due to any given individual are packaged together and analyzed in quantitative genetical terms, in agreement with Hamilton's original viewpoint. Of particular interest is what the heritability via partners' offspring, $\beta_{G'P}$, actually represents. In fact it is proportional to the extent to which partners' offspring bear a phenotypic resemblance to the focal individual.[9] This heritability thus captures Hamilton's insight that, for social evolution, individuals can "reproduce" not only through their direct offspring, but through the offspring of their genetic relatives. The heritability also enables a calculation, in genetic terms, of the value that an individual should place on social partners' offspring, relative to its own (see note 10 and also [Frank, 1998, Frank, 2013]); this supports Hamilton's statement that "the individual will seem to value his neighbours' fitness against his own according to the coefficient of relationship appropriate to that situation" [Hamilton, 1964b].

Dividing equations (8.3) and (8.4) by β_{GP} we can also follow an alternative, quantitative genetics route to deriving variants of Hamilton's rule [Frank, 2013]. The resulting variants correspond to the genotype–phenotype version of Hamilton's rule (HR2 in chapter 7) due to David Queller [Queller, 1985, Queller, 2011]; in one of these variants, derived from the inclusive fitness perspective, assortment can be directly rewritten as the geometric view of relatedness[10] (HR4 in chapter 7). Thus the inclusive fitness approach of equation (8.4) also allows Hamilton's rule with geometric relatedness to be derived; the neighbor-modulated version of equation (8.3) can obscure the importance of relatedness, however, requiring conversion to the equivalent inclusive fitness form in order to make the connection.

Here, as in earlier chapters, simplifying assumptions have been made in showing that inclusive and neighbor-modulated fitness formulations are equivalent. The intention has been to show that inclusive fitness is the quantity that makes sense in evolutionary causal terms. In chapter 10 we will argue that in more complicated scenarios it is still inclusive fitness that natural selection acts on, even when we have trouble analyzing social evolution from an inclusive fitness viewpoint. Having established, at least for sufficiently simple cases, that inclusive fitness makes sense as the quantity that natural selection acts on, we next discuss some details of the maximization metaphor.

8.3 DO INDIVIDUALS ACT TO MAXIMIZE THEIR INCLUSIVE FITNESS?

Hamilton conceived the contribution of inclusive fitness as being an extension to classical fitness that provided a new maximization argument, capable of explaining social behavior. Hamilton saw this as applying to average inclusive fitnesses calculated across classes of individuals, but he also placed emphasis on the importance of the individual, writing in the abstract of his first technical paper on the subject that "species following the model should tend to evolve behaviour such that each organism appears to be attempting to maximise its inclusive fitness" [Hamilton, 1964a].

Andy Gardner has also informally advocated the "individual as maximizing agent" analogy, as discussed within the first chapter in the context of apparent biological design [Gardner, 2009]. The seeming fact that individuals should act as if to maximize their inclusive fitness provides a useful causal perspective on the evolution of behavior and other aspects of organismal phenotype. This perspective also helps disambiguate which form of fitness makes sense in evolutionary terms: inclusive or neighbor-modulated. The analogy favors inclusive fitness since an organism has control over its own behavior, and so may be thought of as making the necessary valuation of own and partners' offspring in deciding whether to act socially. Under the mathematically equivalent neighbor-modulated viewpoint, however, an organism need not have any control over its social partners', potentially correlated, behaviors; hence we cannot think of the organism being selected to maximize neighbor-modulated fitness.

How well has Hamilton's claim for individuals being selected to maximize their inclusive fitness held up? For one thing, as previously discussed, David Queller has pointed out that it is individual behaviors that should be thought of as under selection to maximize inclusive fitness; as noted in chapter 6, apparent paradoxes that arise by considering the individual as an inclusive fitness maximizing agent, such as deciding not to help a dominant relative to reproduce when an individual finds itself in a subordinate role in a cooperatively breeding species, are resolved by considering inclusive fitness maximization for each behavior in turn [Queller, 1996]. So, it seems clear that if individuals do maximize their inclusive fitness, our analysis needs to take place within the constraints that those individuals find themselves behaving under;

this falls under the label of "implicit constraints" on behavior that we discussed in chapter 6.

Noting the lack of formal support for Hamilton's original claim for inclusive fitness as the quantity maximized by natural selection, Alan Grafen, as part of his *formal Darwinism* project [Grafen, 2007a], has analyzed selection on social behaviors [Grafen, 2006a]. Grafen shows that natural selection does maximize inclusive fitness under some assumptions, and crucially his approach begins with neighbor-modulated fitness, but converts it to inclusive fitness in the process. An important restriction in the analysis, however, is that it does not apply to frequency-dependent selection, and therefore does not apply to nonadditive fitness interactions between social partners [Grafen, 2006a].[11] Additionally, the formal Darwinism approach requires there be a plastic relationship between genes and behavior, that can be shaped by evolution [Grafen, 2006a]; yet in our simplest models, behavior is unconditionally expressed. It is beyond the scope of this book to advance beyond Grafen's results, however our intuition that inclusive fitness should still be maximized when social interactions are nonadditive may still be supported by referring back to our first, and simplest, description of evolution under natural selection: the replicator dynamics of chapter 2. Recall from chapter 2 that the asymptotically stable fixed points of selection under the replicator dynamics are those that attract nearby population states, and hence represent the population equilibria that natural selection can result in. With unconditionally expressed behaviors, at a population equilibrium, no individual would benefit from having a different genetic value, given the composition of the population. Each individual's fitness is the maximum it can be, on average and under the current population state. Note, however, that the claim is not that inclusive fitness always increases, but rather that under the equilibria that result from natural selection (if any) no single individual's inclusive fitness would be improved by having a different genetic value and hence a different phenotype, just as in a population of individuals playing the same Nash equilibrium strategy, none could profit by unilaterally changing their strategy. In other words, at population equilibrium, individuals appear to act *as if* they are maximizing their inclusive fitness, given the population they find themselves in and the constraints on their behavior. Samir Okasha and Cedric Paternotte note potential problems with this approach in general, however, as games can be constructed in which the replicator dynamics, and its continuous-trait counterpart *adaptive dynamics*, need not lead populations to occupy such equilibria [Okasha and Paternotte, 2014]. For games in which evolutionary equilibria *are* attainable, it is tempting to imagine that the population frequencies of prosocial and asocial individuals would converge on the mixed-strategy Nash equilibrium that would result if players were able to choose whether to be social according to some behavioral rule; however, Alan Grafen has shown that the pure-strategy and mixed-strategy equilibria need not be the same [Grafen, 1979].

All the foregoing discussion is not to say that the formal Darwinism approach to inclusive fitness [Grafen, 2006a] should not be extensible to the evolution of phenotypes specifying mixed strategies for nonadditive interactions, in which case we would expect the evolved phenotypes to correspond to Nash equilibrium strategies; rather it is to suggest how maximization analogies might be applied to simpler, unconditional behaviors. Laurent Lehmann and François Rousset, however, present a critique of the formal Darwinism project in terms of its relationship to Hamilton's original model, and to analyses using evolutionary game theory; their conclusions are that inclusive fitness changes are fundamental in any maximization analysis, and that population allele-frequency changes under selection look *as if* individuals are changing their behavior to increase their inclusive fitness [Lehmann and Rousset, 2014].[12]

8.4 ULTIMATE CAUSES AND SOCIAL EVOLUTION

Given the explosion of results on social evolution, and in particular, theoretical mechanisms shown to permit the evolution of social behavior, some kind of framework to make sense of the field as a whole seems necessary. In understanding biology, two main levels of explanation exist, first identified by Niko Tinbergen [Tinbergen, 1963]: proximate and ultimate. *Proximate explanations* are mechanistic explanations, explaining *how* a particular biological phenomenon arises in terms of the underlying machinery of the organism's body. *Ultimate explanations* explain *why* biological phenomena occur, in terms of their underlying evolutionary causes. An example of a proximate explanation for flight, to return to an example from chapter 1, would explain how the shape of a wing generates sufficient aerodynamic lift to enable an organism to cover large distances in the air; an ultimate explanation of the same trait would propose reasons why the particular life history of that organism meant it was useful in evolutionary fitness terms for it to fly, for example in order to escape from predators, or to be a more effective predator itself.

Proximate and ultimate approaches to building frameworks for understanding social evolution theory are thus natural approaches to take. In the following we will consider examples of both.

8.4.1 Mechanistic Classification of Social Evolution

We start with a framework for classifying social behavior proposed by Laurent Lehmann and Laurent Keller [Lehmann and Keller, 2006b], which begins with a particular model of social behavior and then proceeds to analyze the different means by which social behavior can experience positive selection within this model. In some senses, the approach is not strictly mechanistic and is similar to the causal frameworks discussed immediately below. However, the approach assumes a particular social interaction structure and all the analysis rests on this assumed structure, and the mechanisms it can represent. Thus, the model presupposes, for example, the possibility for reciprocal cooperation, examines conditions under which reciprocal cooperation can evolve, and then classifies reciprocal cooperation as one route to selection for social behavior. Although the reasoning is circular, this is logically unproblematic; however the assumption that reciprocal cooperation is important in nonhuman animals is itself subject to debate.[13] The more fundamental difference from the causal approaches discussed below is that the analysis starts with a particular model of social interactions, which necessarily gives rise to a number of mechanistic routes to the evolution of social behavior. The question then is whether any biological instance of social behavior exists that corresponds to each of these mechanisms. Lehmann and Keller's approach was well received by the community of inclusive fitness theorists,[14] and represented an important early step in identifying the problems posed by an ever expanding set of models proposing mechanisms by which social behavior can evolve, in classifying existing influential models within a broad framework, and in particular, in emphasizing the importance of the distinction between direct and indirect fitness effects.

While the purpose of their article is broader, encompassing the importance of precise definitions of terminology in the field of social evolution, and the usefulness of the distinction between proximate and ultimate evolutionary explanations, Stuart West, Ashleigh Griffin, and Andy Gardner also offer a combined mechanistic and ultimate taxonomy of explanations for the evolution of social behavior (see [West et al., 2007b, figure 1]).

8.4.2 Causal Frameworks for Social Evolution

8.4.2.1 Exact-Total Analysis

As often repeated, the opening sentence of Fisher's *The Genetical Theory of Natural Selection* [Fisher, 1930] is "Natural selection is not evolution." All the variants of Hamilton's rule we saw earlier in this book have been derived solely from considering the selection term in the Price equation, and assuming the transmission bias term is 0. By allowing for nonzero transmission bias we can arrive at what Steve Frank refers to as an *exact-total* version of Hamilton's rule [Frank, 1997]; rearranging the Price equation (3.3) shows that total evolutionary change in genetic value is positive whenever selection in favor of individuals with larger genetic value outweighs any transmission bias by which offspring tend to inherit smaller genetic values than those of their parents (or vice versa), in which case

$$\text{Cov}(G, W) > -\text{E}(W\Delta G). \tag{8.5}$$

We can then choose our preferred variant of Hamilton's rule, from chapter 7, and use it to replace $\text{Cov}(G, W)$ in the above equation. Various rearrangements can then be made to emphasize important prerequisites for evolutionary change. Frank, for example, prefers an arrangement that highlights the separation between components of direct and indirect fitness, and heritability [Frank, 1997]. An alternative arrangement is suggested in [Marshall, 2011c], with the argument that ultimate causal interpretations of particular mechanisms can be made by writing them in terms of the exact-total condition (8.5), with a particular variant of Hamilton's rule substituted in, and then identifying ultimate causes as corresponding to any change in a component of the resulting equation that results in the condition being satisfied. Marshall presents some examples using this approach, including a reanalysis of results for the evolution of altruism on graphs due to Hisashi Ohtsuki and colleagues [Ohtsuki et al., 2006] in terms of Hamilton's rule with genetic relatedness.[15] Both [Frank, 1997] and [Marshall, 2011c] discuss conditions under which transmission bias may be expected to be nonzero, with regard to culturally inherited traits; Frank presents a "rebellious child" model in which offspring tend to exhibit different behavior to their parents, while Marshall discusses conformity bias in cultural evolution models (see chapter 10), in which individuals are likely to change their behavior to match the dominant behavior in a group they join [Boyd et al., 2011]. Significantly predating such analyses, Sewall Wright had pointed out the potential for social parasitism to result in a positive transmission bias coexisting with negative selection for the parasite ([Wright, 1969], discussed by Michael Wade [Wade, 1985]).[16]

8.4.2.2 Kith, Kin, and Kind

Several variants of Hamilton's rule have already been presented in earlier chapters. David Queller, who derived several of them, has subsequently proposed several refinements of these rules to expose the underlying causes explaining the evolution of social behaviors [Queller, 2011]. Queller separates these refinements into three broad categories: *kin selection*, referring to selection based on standard inclusive fitness interests in family groups,[17] *kith selection*, covering social interactions including manipulation of partners and interspecific mutualisms, and *kind selection*, covering conditional or nonadditive behaviors, such as greenbeard donation.

Starting with kin selection first, Queller derives a version of Hamilton's rule that is intermediate between the variants HR2 and HR3 presented in chapter 7, as [Queller, 2011]

$$\beta_{WP|P'} + \beta_{W'P|P'}\frac{\text{Cov}(G', P)}{\text{Cov}(G, P)} > 0. \tag{8.6}$$

This version of Hamilton's rule combines partial regressions on phenotype, rather than genetic value, with conditional expression of phenotype. This variant can be derived very similarly to the variants seen in chapters 5 and 7,[18] but note that the partial regression on *phenotypes* no longer guarantees the separation condition that the residuals from the regression are uncorrelated with the focal individual's genetic value, and rather simple models can violate this requirement.[19]

Queller treats kind selection by extending condition (8.6) to give [Queller, 2011]

$$\beta_{WP|P'} + \beta_{W'P|P'}\frac{\text{Cov}(G', P)}{\text{Cov}(G, P)} + \beta_{W'(PP')}\frac{\text{Cov}(G', PP')}{\text{Cov}(G, P)} > 0. \tag{8.7}$$

In terms of the variants of Hamilton's rule presented in chapter 7, condition (8.7) is now a combination of aspects of HR1, HR2, and HR3, with the effect of the interaction between focal individual's and partners' phenotypes ($P \times P'$) on the focal individual's fitness replacing the single deviation term D used in HR1.[20] The addition of the interaction term allows the separation condition for the regression of fitnesses on phenotypes to be satisfied, just as the equivalent term did for the independent regressions of fitnesses on genetic values in HR1. As noted in chapter 5, conditional expression of phenotype can be reframed in terms of nonadditive fitness interactions, and vice versa, suggesting that condition (8.7) may introduce unnecessary complication to analysis of a behavior even if it captures all the details of that behavior. Queller interprets kind selection as covering greenbeard traits, as well as reciprocal cooperation, and therefore considers these to be different to standard kin selection. As previously discussed in chapter 7, however, greenbeards can be interpreted as directing fitness benefits to genetic relatives by translation to the geometric view of relatedness (a point Queller also acknowledges [Queller, 2011]), while Sean Nee has previously shown how benefits from reciprocal cooperation can be decomposed, again using the geometric view, into benefits dispersed at random in the population, and synergistic benefits directed only to genetic relatives [Nee, 1989].

Things get really interesting in dealing with kith selection, where Queller takes several approaches to extending the model used to describe a focal individual's fitness. Queller considers behavior by a focal individual that manipulates or influences behavior in social partners, with an incidental effect on the fitness of those partners.[21] Jeremy van Cleeve and Erol Açkay have noted that relatedness need not interact additively with such responsiveness, except in rare cases [van Cleeve and Açkay, 2014], but in the following we shall restrict ourselves to Queller's original, simpler, analysis. Selection for such behaviors in these simpler cases is then shown[22] to be favored when [Queller, 2011]

$$\beta_{WP|P'} + \beta_{WP'|P}\beta_{P'P} > 0, \tag{8.8}$$

where the regression coefficient $\beta_{P'P}$ captures the influence of the focal individual, via their own phenotype, on the phenotype of their social partners. Condition (8.8) shows that personally costly behavior $(\beta_{WP|P'} < 0)$ can arise without genetic relatedness, provided that the individual's behavior modifies that of their social partners, and that modified behavior feeds back on the focal individual's direct fitness; a typical scenario might be that increasing the expression of the partners' social phenotypes $(\beta_{P'P} > 0)$ has positive direct-fitness benefits via feedback $(\beta_{WP'|P} > 0)$ and thus experiences positive selection, but we can also see that if the partners' phenotypes negatively affect an individual's direct fitness $(\beta_{WP'|P} < 0)$ then expression of a phenotype that *decreases* the expression of the social partners' phenotype $(\beta_{P'P} < 0)$ can also experience positive selection. This latter case corresponds to selection for behaviors that reduce selfish behaviors in others, that in turn have incidental harming effects as a by-product. Queller shows that such influencing or manipulation behavior can experience positive selection even when there are incidental effects on social partners' fitness, provided that the feedback from the modification of the social partners' behavior is sufficiently large.[23]

Alternatively, Queller reasons, an individual may benefit by increasing partners' fitness, when that increased fitness translates back to direct fitness benefits for the focal individual. Consider a description of individual fitness in terms of individual's genetic value and social partners' fitness,

$$W = \beta_{WG|W'}G + \beta_{WW'|G}W' + \varepsilon_W. \tag{8.9}$$

Given this description, asking when genetic value is selected to increase in the population $(\mathrm{Cov}(G, W) > 0)$ shows us that the condition is[24]

$$\beta_{WG|W'} + \beta_{WW'|G}\beta_{W'G} > 0. \tag{8.10}$$

This condition shows that an individual can experience selection to increase the fitness of an unrelated social partner, when that social partner's increased fitness feeds back on their own. Importantly, since there is no need for genetic relatedness, the partners may be members of different species; Queller gives the example of a lichen in which extra carbon production by the algal partners might increase the fitness of the fungal partner, which in turn produces more nitrogen that increases the fitness of the algal partner [Queller, 2011]. Thus, under this viewpoint, examples of apparently costly behavior that benefits members of different species can be reinterpreted in terms of direct fitness benefits to members of the focal species arising from their own behavior. We shall return to this point below.

8.4.2.3 Assortment and Heritability

In concluding this chapter, we return to the issue of genotype–phenotype assortment first raised in chapter 7, where we showed that genotype–phenotype assortment (HR2) is equivalent to the geometric view of relatedness (HR4). Steve Frank has argued for the value, in seeking to understand the underlying causes of social evolution, of taking an extended view on Hamilton's rule considering two kinds of relatedness: one between social partners, and one capturing transmission of genetic value via the offspring of social recipients (e.g., [Frank, 2013]). This enlarged view of Hamilton's rule is supported by class-structured models in which some classes never have the opportunity to act socially, and therefore cannot make an indirect fitness contribution to those individuals that *do* behave socially [Frank, 2013]; also supporting the view is the suggestion that genotype–phenotype assortment can arise due to nongenetic correlations, and as such is distinct from genetic relatedness [Frank, 2013], while finally models can be constructed that appear to support the evolution of altruism between species, in which genetic relatedness must be zero [Frank, 1994, Frank, 2013].

In response to these three points indicating the need for an extension of Hamilton's rule, and a distinction between different forms of relatedness, we can make three observations. First, as shown earlier in this chapter, derivations of genetic relatedness either due to correlation between social partners, or due to valuation of others' offspring relative to own through ratios of heritabilities, arrive at the same expression for relatedness (see note 10). Thus whether relatedness describes correlations between equal social partners, or relatedness to offspring of the beneficiaries of social behavior who will never be able to act socially in their turn, the same expression describes both. Second, as shown in chapter 7 (see section 7.2), the genetic geometric view of relatedness underlies genotype–phenotype assortment, and when the latter assortment is positive, the former *must also* be positive; thus phenotypic correlations in social behavior within a species always involve genetic correlations, regardless of the underlying cause. Third, while models of interspecific social behavior apparently can give rise to altruism between species, further analyses have shown that such behavior is frequently vulnerable to modifiers that suppress it [Quickfall et al., 2015], and even when it is stable can be explained by indirect fitness benefits within a focal species [Quickfall et al., 2015, Wyatt et al., 2013]. David Queller's kith selection, discussed immediately above, also provides useful illustrations in this regard; Queller shows how initially costly behavior of a focal species, that potentially increases the fitness of partner species, can be selected for when the benefits to the partner species feed back as sufficiently large direct fitness benefits to the focal species. When such return benefits do not accrue directly to the focal social individual, they must instead accrue to their genetic relatives within the same species if that social behavior is to experience positive selection. This is actually likely to be the situation in Queller's lichen example quoted above, since the benefits of increased nitrogen production by members of the fungal partner species are not received solely by a single carbon-providing alga. Providing theoretical support for this view, Kevin Foster and Tom Wenseleers show that high within-species relatedness facilitates costly donation between species, while low within-species relatedness undermines it [Foster and Wenseleers, 2006].

8.5   Summary

The analyses presented in this chapter suggest that, despite its predictive equivalence with other apparently different selective frameworks, such as multilevel selection, inclusive fitness comes out as the only causal explanation for observed social evolution, and also can be cast as the quantity that natural selection maximizes. The Price equation is particularly well suited to providing causal explanations for particular mechanisms proposed to allow the evolution of social behavior, both within species and between species. These analyses show that, within species, genetic relatedness is always the fundamental quantity for describing both correlations between social partners, and heritability of traits via the offspring of the recipients of social behavior. Furthermore, when costly social behavior affecting another species occurs, either direct fitness benefits or inclusive fitness benefits within the focal species must explain the evolution of that behavior.

CHAPTER NINE

What Is Fitness?

9.1 INTRODUCTION

Throughout this book fecundity, or production of offspring, has been treated as synonymous with evolutionary fitness. While this is valid for the relatively simple models already analyzed in the book, in more biologically realistic scenarios fecundity and fitness need not be the same at all, and can give qualitatively different predictions when used within inclusive fitness analyses such as versions of Hamilton's rule. In this chapter we shall give a brief treatment of what evolutionary fitness is properly understood as being, and briefly review proposals on how to deal with fitness appropriately within inclusive fitness theory. Since these are complicated issues, the treatment is brief and illustrative, but should be sufficient to raise awareness of some of the key issues. As we will see, it is of particular importance to recognize that payoffs in evolutionary games do not necessarily correspond to the true fitness changes that inclusive fitness keeps track of.

9.2 HALDANE'S DILEMMA

As discussed in chapter 1, J.B.S. Haldane considered the hypothetical question of when an individual should risk their life to save that of a drowning relative.[1] Jumping into a river to save another appears to be altruistic behavior, that we should therefore be able to analyze with inclusive fitness theory and Hamilton's rule. Yet what are the benefits and costs associated with the social act, that we require in order to apply inclusive fitness theory? Clearly since lives are either saved or lost, benefits and costs must relate to how "valuable" those lives are in evolutionary terms. So, we can formalize *Haldane's dilemma*[2] by writing the fitnesses of the individual on the riverbank and in the river respectively as

$$\overset{\text{expected fitness if aid not given}}{W = (1 - P) \overbrace{V} + P \underbrace{\mu V}_{\text{expected fitness if aid given}}} \tag{9.1}$$

and

$$\overset{\text{expected fitness if aid not given}}{W' = (1 - P) \overbrace{V\mu'} + P \underbrace{V}_{\text{expected fitness if aid given}}}, \tag{9.2}$$

where P represents the focal individual's behavior (0 when the individual does nothing to help, and 1 when they jump in to attempt rescue), μ represents their probability of surviving a rescue attempt, and μ' represents the drowning individual's probability of surviving if they receive no aid; note that for simplicity we assume that a rescue attempt will always be successful even if the rescuer drowns in the process. What does V represent? It is the value to an individual, in evolutionary terms, of their life. The fitness equations above highlight that the risk of death[3] can make fitness effects all-or-nothing; if an individual survives then the value to them of their life is the evolutionary fitness they will enjoy in the future, whereas if they die then their future evolutionary contribution is 0 since they can no longer reproduce either directly or indirectly. Leaving further discussion of what V is to the following section, we can work out a condition for the behavior of jumping into rivers to save drowning individuals to experience positive selection as[4]

$$\frac{\text{Cov}(G', P)}{\text{Cov}(G, P)} > \overset{\text{probability altruist dies giving aid}}{\frac{\overbrace{1 - \mu}}{\underbrace{1 - \mu'}_{\text{probability recipient dies without aid}}}}. \tag{9.3}$$

This condition corresponds to Hamilton's rule in its $r > c/b$ form, meaning that the cost in Hamilton's rule is proportional to $1 - \mu$ (the probability that an altruist dies in attempting a rescue) while the benefit to the individual in the river is proportional to $1 - \mu'$ (the probability that they would die if unaided).[5] The covariance ratio on the left-hand side of condition (9.3) is relatedness, but it is interesting to briefly note that this ratio has been derived as a ratio of heritabilities as suggested in chapter 8, since we assumed that drowning individuals never have the opportunity to repay the favor of being rescued (see note 4), and to recall that this ratio can be expressed in genetic terms as the geometric view of relatedness (see chapter 7, note 7).

9.3 Reproductive Value and Class Structure

Our consideration of Haldane's dilemma above assumed that, in some sense, the lives of the potential rescuer and the drowning individual are equivalent. However, as Hamilton noted,

> Haldane does not discuss the question which his remarks raise of whether a gene lost in an adult is worth more or less than a gene lost in a child. However, this touches an aspect of the biological accounting of risks which together with the whole problem of the altruism involved in parental care is best reserved for separate discussion. [Hamilton, 1964b]

We can easily relax our previous assumption to derive a more general version of the condition for rescuing to be favored in Haldane's dilemma, as

$$\frac{\beta_{G'P}}{\beta_{GP}} > \frac{V(1-\mu)}{V'(1-\mu')}, \tag{9.4}$$

where V is now the "value" of the focal individual's life, and V' is the "value" of the drowning individual's life. But what *is* the value of an individual's life? The proper definition of fitness has, since the work of R. A. Fisher, been recognized to be a measure of the success of an individual in contributing their genes to the population gene pool in the distant future. Fisher described this as individuals' *reproductive value* and, taking an actuarial approach to evolutionary fitness, considered the issue broached by Hamilton above, of the reproductive value of individuals of different ages, writing,

> To what extent will persons of this age, on average, contribute to the ancestry of future generations? The question is of some interest, as the direct action of Natural Selection must be proportional to this contribution. [Fisher, 1930]

As shown in figure 9.1, Fisher calculated reproductive values from real demographic data on birth and death rates according to age; using the analogy of compound interest, offspring have their own offspring, and so on ad infinitum, just as interest on interest accrues over time, indefinitely if the loan is not paid off.

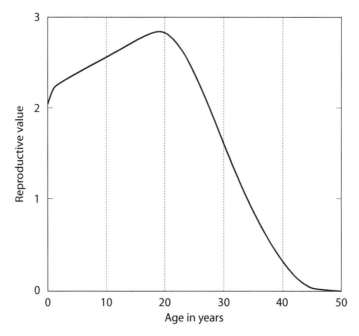

Figure 9.1: Reproductive value of Australian women. The reproductive value for female persons calculated from the birth and death rates current in the Commonwealth of Australia about 1911. Figure redrawn and legend reproduced from [Fisher, 1930].

As figure 9.1 shows, real biological populations typically are *class structured*, with the expected reproductive value of an individual often depending strongly on the class they currently occupy. Figure 9.1 also answers Hamilton's rhetorical question from above; due to their much higher reproductive value, at least for Australian women in the early twentieth century, individuals of around 19 years of age should be more worthwhile saving from the river, followed by children, followed by women in their late twenties and onwards. In particular, the rapid reduction in reproductive value towards 0 illustrated in figure 9.1, likely to be typical in many species, seems to suggest that altruistic behavior should be much more easily selected for among older, postreproductive individuals. If the reproductive value V of a potential rescuer was close to 0, then according to condition (9.4) they should risk their lives saving even very distant genetic relatives.[6] Several reasons might explain why we do not typically hear of sexagenarian women jumping into rivers to save drowning children, of course. First, the benefit to their inclusive fitness from not risking their life, via helping their offspring and grand-offspring (e.g., [Williams, 1957, Lahdenperä et al., 2004]), would presumably be much greater than the inclusive fitness benefit from potentially saving the life of a distant genetic relative. Second, the risk of drowning themselves (the cost in Hamilton's rule) may be increased due to their age, and the improvement in the survival chances of the drowning individual (the benefit in Hamilton's rule) may also be reduced.

Two points are worth noting about reproductive value. First, since true reproductive value is a measure of an individual's contribution to ancestry of the population far into the future, it can never be precisely measured empirically; rather, the reproductive value of classes can be estimated from demographic data, as in the example of figure 9.1, while the reproductive value of individuals can be estimated from proxies such as the number of offspring surviving to reproductive age, or the number of grand-offspring, or even aspects of state as discussed below. In models, however, reproductive value can be calculated or well approximated in certain cases, as discussed below. Second, while Fisher considered reproductive value as a function of age, in fact age is just one aspect of an individual's state that can affect their reproductive value; many, many other aspects of individual state, such as body mass, possession of a territory, and so on, can affect reproductive value [Houston and McNamara, 1999]. Peter Taylor described how inclusive fitness models can be constructed to deal with class-structured populations and reproductive value [Taylor, 1990], and experimentalists have also tackled the problems associated with measuring inclusive fitness in class-structured populations, such as in age-structured populations (e.g., [Oli, 2003]).

9.4 Fitness, Fecundity, and Payoffs

Costs and benefits can be expressed in a number of ways: as payoffs that are proportional to reproductive success, as changes in production of offspring, or as changes in evolutionary fitness. As discussed above, fitness is the relevant quantity in predicting evolution in the long term. Furthermore, the b and c of Hamilton's rule (4.1) properly represent evolutionary fitness. Yet fitness need not be the same as payoffs, or even the same as fecundity. In the following we shall see examples of when they are different, and how inclusive fitness analyses can be performed properly in such situations.

9.4.1 Is Altruism Possible in Viscous Populations?

In presenting inclusive fitness theory, Hamilton considered mechanisms by which the relatedness necessary for altruism to evolve might arise. One influential proposal he made, that has dominated ecological hypotheses for the evolution of animal societies, was that *viscous populations* in which individuals disperse only a short distance from their place of birth, should create the necessary population structure for positive selection on altruism. As he wrote in his original papers,

> With many natural populations it must happen that an individual forms the centre of an actual local concentration of his relatives which is due to a general inability or disinclination of the organisms to move far from their places of birth. In such a population, which we may provisionally term "viscous," the present form of selection may apply fairly well to genes which affect vagrancy. It follows… that over a range of different species we would expect to find giving-traits commonest and most highly developed in the species with the most viscous populations whereas uninhibited competition should characterise species with the most freely mixing populations. [Hamilton, 1964a]

Subsequent computational work by David Sloan Wilson and colleagues [Wilson et al., 1992], and analytic results by Peter Taylor [Taylor, 1992a, Taylor, 1992b] appeared to show that in *inelastic* viscous populations, which cannot grow, local altruism towards relatives experiences negative selection. This appears to run counter to Hamilton's suggestions and to a simple application of Hamilton's rule, in which locally interacting individuals are indeed related, but why? The solution to the problem was to recognize that *all* of the effects of social behavior in such populations had not been accounted for when applying Hamilton's rule; while altruism towards relatives did indeed increase their reproductive potential, given the inelastic nature of the population, increased reproduction by some individuals must always be offset by reduced reproduction by others. In the models of Wilson et al. [Wilson et al., 1992] and Taylor [Taylor, 1992a, Taylor, 1992b], the population structure is such that the individuals forced to reduce their reproduction are related to altruistic donors to exactly the same degree as the recipients of altruism are related to those donors. In summary, the scale of local interaction and the scale of local competition for reproduction are identical and cancel each other exactly, with the net result that altruism does not experience positive selection.

Two main approaches to accounting for all the fitness effects of a social behavior in Hamilton's rule are practicable. The first is to recalculate the benefits arising from social behavior, to account for possible increases in competition that may also result. The second is to change the way in which relatedness is calculated, to account not only for relatedness to recipients of altruism, but also to individuals adversely effected as a result of altruism, through increased competition experienced by their own offspring. These approaches are inspired by models of completely inelastic populations. Although not considered here, it can also be important to take account of social traits that increase local environmental carrying capacity, such as production of public goods in bacteria, as shown by Thomas Platt and James Bever [Platt and Bever, 2009]. Other aspects of life history and demography can also support altruism in viscous populations, as reviewed by Angela Yeh and Andy Gardner [Yeh and Gardner, 2012].

9.4.2 Recalculating Benefits to Account for Competition

Starting first with the recalculation of benefits arising from altruism to account also for increases in competition for reproduction, Steve Frank showed[7] that an extended version of Hamilton's rule,

$$r(b - a(b - c)) - c > 0, \tag{9.5}$$

determines whether altruism experiences positive selection [Frank, 1998]. In this extended Hamilton's rule, b and c are the direct costs and benefits of altruism, in terms of immediate changes in reproductive potential, specified by the payoffs of the donation game (table 2.1). Together with the scale of competition over reproduction, a, these can be used to calculate a corrected benefit B to replace that originally used in the simplest form of Hamilton's rule (4.1). The above condition can thus be written as Hamilton's rule with a properly calculated benefit parameter:

$$rB - c > 0. \tag{9.6}$$

When competition is global ($a = 0$), as would be the case if individuals expressed social behaviors towards local neighbors, before dispersing globally in order to reproduce, then condition (9.6) simplifies to the original version of Hamilton's rule with benefit defined as immediate payoff b received due to altruism (4.1), and this version accurately predicts when altruism experiences positive selection, as shown in figure 9.2. As also shown in figure 9.2, however, as the scale of competition decreases (a increases), so altruism requires higher and higher levels of relatedness r in order to evolve until, when competition for reproduction occurs over the same scale at which social interactions take place ($a = 1$), altruism and competition exactly cancel out and altruism can no longer evolve.[8]

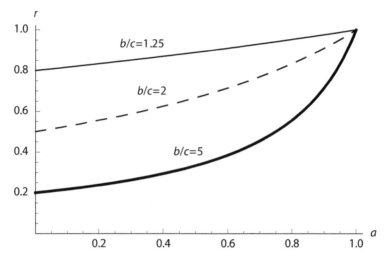

Figure 9.2: Threshold relatedness with varying scales of competition. When competition for reproduction is global ($a = 0$), then the threshold value of relatedness r required for altruism to evolve is predicted by Hamilton's rule considering only the immediate benefits of altruism, b. When competition for reproduction is entirely local, and at the same scale as altruistic interactions take place ($a = 1$), then altruism can arise only when relatedness r exceeds 1, regardless of the ratio of the immediate benefits of altruism b to its cost c. Figure redrawn from [West et al., 2001].

Stuart West and colleagues used the extended Hamilton's rule (9.6) to suggest why the level of fighting between male fig wasps varies not with their relatedness, as a simple application of Hamilton's rule would suggest, but rather varies according to a proxy for intensity of local competition: local female density [West et al., 2001].

9.4.3 Compensated Relatedness

As an alternative to modifying the benefit in Hamilton's rule to account for competition, relatedness instead can be recalculated to take account of competitive effects. Specifically, relatedness should account for not only the relatedness between potential altruist and potential recipient, but also the relatedness between the altruist and those individuals that will experience increased competition due to the increased reproduction of the recipient. For every b units of increased reproduction by the recipient of altruism, therefore, in an inelastic population b units of reproduction will need to be lost by other individuals, so Hamilton's rule accounting for *all* effects of a social behavior on genetic relatives becomes

$$rb - r_e b - c > 0, \qquad (9.7)$$

where r_e is the relatedness of the focal individual to those individuals whose offspring are displaced [Queller, 1994, Grafen and Archetti, 2008]. This condition can easily be rearranged as

$$r_c b - c > 0, \qquad (9.8)$$

where $r_c = r - r_e$ is a recalculated version of relatedness that can be referred to as *compensated relatedness* [Grafen and Archetti, 2008]. This approach to dealing with competition effects can be traced back to David Queller [Queller, 1994], who derived a geometric version of compensated relatedness.[9] Alan Grafen and Marco Archetti, building on work by Peter Taylor, Troy Day, and Geoff Wild [Taylor et al., 2007a], have applied this approach to spatial inelastic populations, and been able to provide explanations for previous results on how different population structures and selective regimes differ in the selective pressures experienced by altruism [Grafen and Archetti, 2008]. Grafen and Archetti's results explain some apparent contradictions; for example, in some simple spatial models, selective regimes in which individuals are selected to die at random, and then neighbors compete to replace them, lead to the evolution of altruism whereas, without any change in payoffs or population structure, if individuals reproduce first then replace random neighbors, altruism does not evolve [Ohtsuki et al., 2006, Ohtsuki and Nowak, 2006, Lehmann et al., 2007, Grafen, 2007b]. Martin Nowak, Corina Tarnita, and E. O. Wilson [Nowak et al., 2010] subsequently used this result as further evidence for the inadequacy of inclusive fitness theory to explain the evolution of altruism. However, Grafen and Archetti had showed, as others before them had, that

> one cannot simply construct a model and assume that Hamilton's rule applies to parameters one has arbitrarily labelled "*r*," "*b*," and "*c*," any more than a physicist would construct her own model of simple mechanics and expect Newton's second law "$F = ma$" to hold with arbitrarily labelled parameters "F," "m," and "a." [Grafen, 2006b]

9.4.4   Deriving Fitness Costs, Benefits, and Optimal Behavioral Rules

We have now seen how Hamilton's rule needs to be correctly applied in situations where there is competition over reproduction. The versions of Hamilton's rule we have seen for this purpose (conditions (9.6) and (9.8)) calculate benefits and costs of altruism in terms of effects on offspring numbers, or fecundity. As noted at the start of this section, however, the b and c in Hamilton's rule are really changes in evolutionary fitness, and these need not correspond exactly to changes in fecundity. In asocial situations, for example, John McNamara and colleagues have shown how behavior that can seem suboptimal, in terms of maximizing number of surviving offspring, is actually optimal in terms of maximizing long-term descendants [McNamara et al., 2011]. Fundamentally, such results rest on the fact that not only the state (class) of individuals may be variable, but also the state of their offspring may be variable. If we allow our modeled individuals to base their reproductive decisions, including those regarding investment in social behavior, on state (e.g., [McNamara et al., 1994, Day and Taylor, 1997][10]) we should thus consider the evolution of state-dependent behavioral rules, rather than simply the evolution of behavior in a given state [McNamara and Houston, 1996, Houston and McNamara, 1999]. The long-term growth-rate formulation of fitnesses has been applied to simple models in social evolution theory (e.g., [Gardner et al., 2007b]). In general, however, the problem of deriving fitness is complicated by demographic factors such as spatial distribution of the population and variability of the environment [Metz et al., 1992], leaving the subject well beyond the scope of an introductory textbook.

9.5 SUMMARY

In this chapter we have considered the subtleties in applying Hamilton's rule to more complicated social scenarios, which arise when fitness is not correctly interpreted. We have shown that payoffs from interactions, or even changes in immediate offspring numbers, need not correspond to the correct evolutionary definition of fitness required by inclusive fitness theory, in terms of long-term contribution to future generations. We have summarized the problem of considering individuals having different states, and hence different reproductive values. We have considered how to account for the effects of increased competition that can arise from social behaviors when population size is limited; this can be achieved either by recalculating the benefit term in Hamilton's rule, or the relatedness. Finally we have also considered the problems associated with calculating fitness in environments that are inhomogenous, and also mentioned the application of inclusive fitness theory to calculating optimal state-dependent life-history rules.

CHAPTER TEN

Evidence, Other Approaches, and Further Topics

10.1 INTRODUCTION

The previous chapters of this book have attempted to explain the genesis, the logic, and the generality of social evolution theory. In particular, a central aim of the book has been to demonstrate that William D. Hamilton's inclusive fitness theory provides the necessary generalization of classical Darwin–Wallace–Fisher fitness, which allows us to provide evolutionary explanations of the many social behaviors we observe in the natural world. Since the book has focussed on the theory, on evolutionary causes, and on explaining these in a relatively simple and accessible manner, a great many important topics in inclusive fitness theory have been mentioned only in passing, or even omitted altogether. In this concluding chapter we will attempt to rectify this deficit, drawing attention to the tremendous advances in both empirical prediction from, and testing of, Hamilton's theory, the subtleties of its theoretical development, and its potential for application in ever more aspects of the life sciences.

10.2 EMPIRICAL SUPPORT FOR INCLUSIVE FITNESS THEORY

The empirical support for inclusive fitness theory has recently been criticized as being "meagre."[1] This is surprising, given that inclusive fitness theory is among the most extensively tested and verified theories in the biological sciences (e.g., [Abbot et al., 2011, Bourke, 2014]). One aspect of the confusion may be that

115

inclusive fitness theory is frequently conflated with the *haplodiploidy hypothesis*, advanced by Hamilton, that the haplodiploid genetics of the Hymenoptera could explain the examples of eusociality that they provide. Here we briefly discuss empirical support for inclusive fitness theory, which spans a broad range of species, phyla, and kingdoms. We also discuss the haplodiploidy hypothesis and its modern replacement: the *monogamy hypothesis*.

10.2.1 Microbial Altruism

We start our empirical review with the simplest organisms studied: the microbes, although the capacity for social behavior in such organisms attracted recognition and investigation only comparatively recently [Crespi, 2001, West et al., 2007a, Foster, 2011]. In fact, we have already seen examples of microbial social behavior in chapter 1, in stalk formation by *Dictyostelium* species (figure 1.4), in production of iron-scavenging siderophores (figure 1.5), and in production of pyocins by bacteria (figure 1.6). Any study of social behavior needs to consider whether that behavior is costly or beneficial for the individual expressing it, and whether it is costly or beneficial for other individuals (table 2.3). Any application of inclusive fitness theory further requires us to consider the genetic relatedness between actors expressing behaviors, and other individuals affected by them.

As discussed in chapter 1, stalk formation by *Dictyostelium* amoebae is clearly altruistic since members of the stalk die, without further reproduction, in order to facilitate the reproduction of members of the fruiting body. The stalk is thus a kind of ad hoc somatic material, in a primitive multicellular body. So what role does genetic relatedness play in this mode of reproduction? Joan Strassmann, Yong Zhu, and David Queller showed that in *Dictyostelium discoideum*, where stalks and fruiting bodies can be formed by genetically different clones, these clones can "cheat" by forming less than their fair share of the sterile stalk, thereby gaining greater representation and increased relative reproduction in the fruiting body [Strassmann et al., 2000]. In *Dictyostelium purpureum*, Natasha Mehdiabadi and colleagues showed pronounced *genetic kin recognition*; at the precursor life-cycle stage to formation of the stalk and fruiting body, formation of the slug, different genetic clones assorted so that they formed genetically homogenous groups not vulnerable to the manipulation observed in *D. discoideum* [Mehdiabadi et al., 2006]. Owen Gilbert and colleagues measured relatedness in wild fruiting bodies of *D. discoideum*, and showed that this was high enough to select against cheater mutants that do not contribute to stalk formation [Gilbert et al., 2007]. These observations, genetic kin recognition to avoid altruism being directed to nongenetic relatives [Hamilton, 1964b, Hamilton, 1971], and favoring genetic relatives over nonrelatives when the opportunity arises [Hamilton, 1964a, Hamilton, 1964b], are both standard predictions of inclusive fitness theory. The theory of genetic kin recognition is discussed further below.

In chapter 1 we were also introduced to production of iron-scavenging siderophores by bacteria. As shown in figure 1.5, production of the siderophore pyoverdin is individually costly but benefits other members of a population, and therefore classifies as altruistic behavior (table 2.3). As Ashleigh Griffin, Stuart West, and Angus Buckling showed, relatedness within populations of *Pseudomonas aeruginosa* leads to increased levels of pyoverdin production, but only when competition between bacteria occurs over larger scales [Griffin et al., 2004]. These results are in agreement with inclusive fitness theory, when extended as described in chapter 9 to account for competition effects. Interestingly, in commenting on the results, David Queller noted that as well as manipulating relatedness and competition within bacterial populations, the experimental design used by Griffin and colleagues also could be interpreted as a group selection process [Queller, 2004]. This design thus provides an experimental example of the theoretical results, presented in chapter 4, on the equivalence of inclusive fitness and group (multilevel) selection perspectives.[2]

Other examples of microbial social behavior abound; for example bacterial greenbeard traits have been discovered, such as *FLO1* discussed in chapter 6 [Smukalla et al., 2008]. Microbial species are rapidly becoming established as model systems for the empirical testing of social evolution theory, given the ease with which they can be maintained in the lab, and their short generation time.

10.2.2 Help in Cooperative Breeders

We next consider vertebrate cooperative breeders. Cooperative breeders, first discussed in chapter 1, form small groups in which reproduction is completely or largely monopolized by a dominant breeding individual or pair [Crespi and Yanega, 1995]. Subordinate individuals apparently contribute to raising the offspring of the dominants in the group, for example through foraging for them, or "babysitting" them. Although vertebrate generation times are much longer than microbial ones, long-term studies of species such as the meerkat (*Suricata suricatta*, figure 10.1) and the long-tailed tit (*Aegithalos caudatus*) have been able to study the indirect and direct fitness benefits of social behaviors in these populations. Such studies have revealed diverse patterns underlying social behavior.

Figure 10.1: A meerkat sentinel (*Suricata suricatta*) engaged in *raised guarding* will emit an alarm call, warning other group members when a predator is nearby. Hamilton identified alarm calling as a possible social behavior that might be explained by benefits to genetic relatives, given a sufficiently small cost in terms of increased predation risk for the calling individual [Hamilton, 1964a]. Subsequent investigation however has shown that guarding (and calling) meerkats experience no additional risk of predation [Clutton-Brock et al., 1999], which theoretically will favor alarm calling even when group members are very distantly related. Photograph by the author.

In meerkats, for example, while some behaviors such as alarm calling may be explained in inclusive fitness terms by negligible direct fitness costs (figure 10.1), performance of individually costly behaviors has been shown

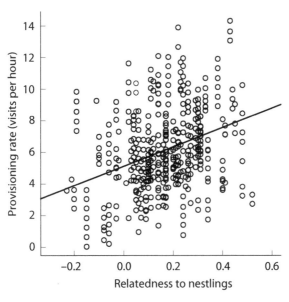

Figure 10.2: Rate of provisioning by long-tailed tit helpers (*Aegithalos caudatus*). Provisioning rate is plotted as a function of genetic relatedness, estimated using techniques based on the geometric view of relatedness (see chapter 7), for 14 years of observations. The regression line is produced by a linear mixed-effect model of the data. Figure redrawn from [Nam et al., 2010].

not to correlate with relatedness between helpers and the offspring they are aiding [Clutton-Brock et al., 1998, Clutton-Brock et al., 2000, Clutton-Brock et al., 2001], although possible exceptions to this pattern include allolactation [MacLeod et al., 2013]. Motivated by observations such as equal contributions to cooperation by unrelated immigrants and individuals born within the group, the *group augmentation hypothesis*, formalized by Hanna Kokko, Rufus Johnstone, and Tim Clutton-Brock, proposes that in some cooperatively breeding species, group members may enjoy increased survivorship, and hence prospects for attaining dominant reproductive status, by cooperating to increase the size of their group [Kokko et al., 2001].[3]

In long-tailed tits, on the other hand, Jessica Meade and Ben Hatchwell found no evidence for direct fitness benefits for helpers [Meade and Hatchwell, 2010], suggesting an important role for indirect fitness in explaining helping behavior in such species. At the same time, Ki-Baek Nam and colleagues showed that helping in the same species positively correlates with relatedness between helpers and nestlings (figure 10.2), although aid by unrelated helpers still occurs

[Nam et al., 2010]. Furthermore, Andy Russell and Ben Hatchwell have shown that unsuccessful breeders that become helpers favor helping at the nest of related breeders [Russell and Hatchwell, 2001], while Ben Hatchwell, Philippa Gullett, and Mark Adams conclude that Hamilton's rule explains the occurrence of helping in long-tailed tits [Hatchwell et al., 2014]. Important differences in life history between long-tailed tits and meerkats are likely to explain the different roles of direct and indirect fitness effects in the two [Clutton-Brock, 2002, Hatchwell, 2009], and comparative analysis has shown that the magnitude of fitness benefits from helping correlates with the extent to which kin recognition occurs within species [Griffin and West, 2003].

10.2.3 Reproductive Restraint in Eusocial Species

The social Hymenoptera (including species of ants, bees, and wasps) are perhaps the best known example of altruism, through the extreme reproductive division of labor they exhibit. The social Hymenoptera are not the only eusocial species,[4] however, nor is individual reproductive restraint always explained by inclusive fitness benefits in such species. We will consider the phylogenetic distribution of hymenopteran eusociality and its consequences in the next section. Here we briefly consider other routes, apart from individual altruistic restraint, by which limited reproduction by workers, or by subordinate queens, may arise or be maintained.

10.2.3.1   Worker Policing

It may seem that altruism between genetic relatives is a sufficient explanation of the reproductive restraint apparently exercised by workers in many social insect species, such as the honeybees. Indeed, as discussed below, Hamilton initially proposed that the peculiar haplodiploid genetic system of the Hymenoptera, of which the honeybees (*Apis mellifera*) are a member, could explain why daughter workers in a colony would be willing, in inclusive fitness terms, to raise the offspring of their mother rather than attempt to reproduce themselves. Interestingly, however, the situation is more complicated. Under haplodiploid genetics, females are diploid while males are haploid, hence unmated females such as workers can still produce males if they retain functioning ovaries. As described below, Hamilton reasoned that high genetic relatedness between workers, arising from haplodiploidy and single queen matings, should favor reproductive restraint by workers [Hamilton, 1964b]. When queens mate multiple times, average relatedness between workers in the colony decreases, which would seem to suggest that worker reproduction should be favored. However, the opposite pattern is observed among the Hymenoptera, as summarized by Francis Ratnieks [Ratnieks, 1988]. Ratnieks constructed a model based on earlier suggestions by Tom Seeley and others [Seeley, 1985] that honeybee workers might selfishly reproduce when queens mate once, but control the reproduction of each other when queens have multiple matings. Such behavior would arise due to conflicting preferences over who produces the colony's sexual males. Under single matings, relatedness patterns mean that workers value their own sons first, then their sister workers' sons, then those of the queen; with multiple mating, however, while the workers still value their own sons first, the preference for queen's versus sisters' sons becomes reversed. This creates a selective pressure for workers to control each others' reproduction, for example through consuming eggs not laid by the queen. Ratnieks's model of *worker policing theory* showed how such a "policing" allele could invade and be stable in a population with reduced relatedness. Worker policing theory has received extensive empirical validation; for example Ratnieks and Kirk Visscher showed discrimination of worker-laid eggs in the honeybee [Ratnieks and Visscher, 1989], and Tom Wenseleers and Ratnieks presented a comparative analysis of 10 hymenopteran species showing both of the relationships predicted by the theory: a positive correlation between within-colony relatedness and worker reproduction, and a negative correlation between effectiveness of policing and worker reproduction (figure 10.3, [Wenseleers and Ratnieks, 2006b]; see also [Wenseleers and Ratnieks, 2006a]). Since multiple mating appears to be a derived characteristic of eusocial species (figure 10.4), policing is thus seen as an evolutionary route to maintaining eusociality as relatedness becomes reduced. Worker policing is also observed in species with lower relatedness, such as the bumblebee *Bombus terrestris*, where it seems to have evolved through selfish competition between workers over reproduction [Zanette et al., 2012].

10.2.3.2   Direct Fitness Benefits in Queen Associations

Earlier in this chapter we saw how, in cooperatively breeding mammals, direct fitness benefits seem necessary to explain the lack of correlation between genetic relatedness to offspring, and incidence of cooperative behavior by helpers. Although relatedness explains patterns of reproduction in most eusocial insect species, instances in which individuals provide aid to nonrelatives can be found, such as in the paper wasp (*Polistes dominulus*). Ellouise Leadbeater and colleagues showed that in single-season multifoundress associations, in which a substantial proportion of associations involve unrelated mated females, the contribution to foundation and

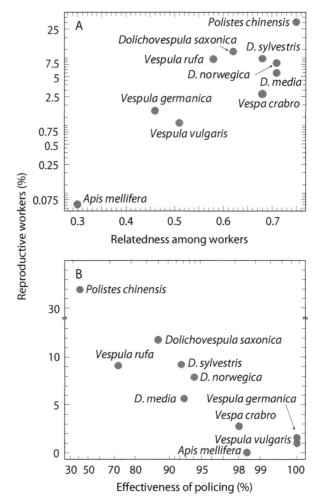

Figure 10.3: A comparative analysis of 10 hymenopteran species supports the predictions of worker policing theory. As predicted by the theory, increased relatedness results in increased attempts at worker reproduction, as workers value the rearing of the queen's sons lowest of all, whereas when relatedness is lower, workers should prefer to rear the queen's sons over those of other workers, and should therefore attempt to restrict the reproduction of other workers (panel a). Worker attempts at reproduction also decrease with the increased effectiveness of policing behavior, through identifying and destroying worker-laid eggs (panel b). Figure redrawn from [Wenseleers and Ratnieks, 2006b].

maintenance of a nest by unrelated subordinate females can be explained both by increased direct reproduction compared to founding a nest alone, even when occupying a subordinate role, and also by the potential for increased direct reproduction through inheriting a nest when the dominant female dies [Leadbeater et al., 2011].[5] Related subordinate foundresses also gain indirect fitness benefits from helping, of course. Leadbeater et al.'s results on direct fitness benefits for unrelated individuals illustrate the fundamental point that social behavior need not be explained only by fitness benefits to genetic relatives, even though inclusive fitness is essential to explain genuine instances of altruism [Griffin and West, 2002].

10.2.4 The Evolution of Eusociality and Cooperative Breeding

The problem of explaining the evolution of the eusocial insects, as discussed in chapter 1, dates back to Darwin, who explained it in terms of selection acting at the colony level [Darwin, 1859, Ratnieks et al., 2010]. With the advent of Hamilton's theory, approaches to understanding eusociality in terms of individual inclusive fitness advantage became possible. In particular Hamilton's original papers [Hamilton, 1964b] introduced an influential hypothesis that many, to this day, erroneously consider as synonymous with his more general theory.

10.2.4.1   The Haplodiploidy Hypothesis

In introducing inclusive fitness theory, Hamilton also made a proposal that the origin of eusociality, at least in the Hymenoptera, could be explained by their haplodiploid genetic system [Hamilton, 1964a]. Noting that the daughters of a singly mated female are more related to their sisters than they are to their own offspring, Hamilton proposed that daughters could have been selected, on the basis of inclusive fitness benefits, to remain with their mother and help raise her offspring, rather than attempt to mate and found their own colony. Early theoretical developments raised issues with the haplodiploidy hypothesis, however; Robert Trivers and Hope Hare noted that, if the colony sex ratio is equal, the increased relatedness of females to their sisters is exactly canceled by their decreased relatedness to brothers,[6] leading Trivers and Hare to refine Hamilton's proposal to be that the increased relatedness of daughter workers to sisters supports the evolution of eusociality only if they can produce males themselves, or act to produce a female-biased colony sex ratio [Trivers and Hare, 1976]; Trivers and Hare presented empirical evidence for their theoretical prediction of 3:1 ratios of females to males in eusocial insect species. Developing the theoretical analysis further, Robin Craig next showed that if colony sex ratios are female biased as the Trivers–Hare theory predicts they should be [Trivers and Hare, 1976], then males are rarer and consequently more valuable in reproductive value terms, so again the

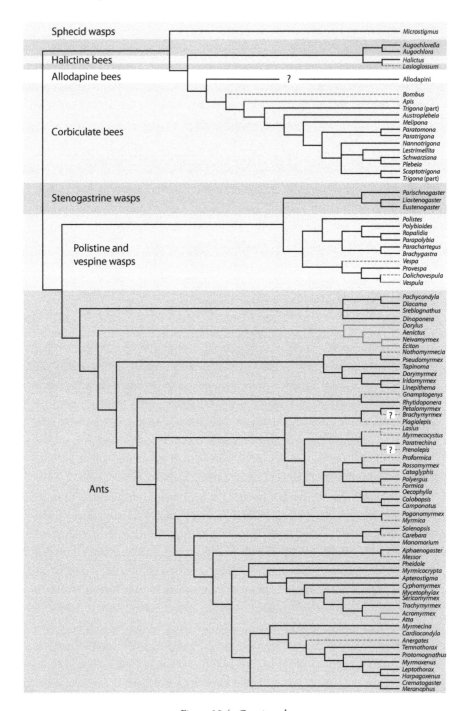

Figure 10.4: Continued.

differences in relatedness between male and female siblings are canceled, leading Craig to conclude that Trivers and Hare's empirical demonstrations of 3:1 sex ratios were not evidence for the importance of haplodiploidy and kin selection in the evolution of eusociality [Craig, 1979].

Trivers and Hare also proposed the importance of *split sex ratios*, in which different colonies exhibit different sex ratios, as part of the evolutionary route to eusociality [Trivers and Hare, 1976]; detailed theoretical investigation by Andy Gardner, João Alpedrinha, and Stuart West has suggested that intercolony sex-ratio variation has played a negligible role in the evolution of eusociality, however [Gardner et al., 2012, Alpedrinha et al., 2013, Alpedrinha et al., 2014].

Empirically, the haplodiploidy hypothesis seems not to hold up either. Although many of the known eusocial species do have haplodiploid genetic systems, far from all of the Hymenoptera are eusocial (figure 10.4). Furthermore, eusocial species with diploid genetic systems, such as the termites, also exist.[7] Haplodiploid genetics have also been shown not to explain variation in the sex ratio of helpers in eusocial species, both empirically as well as for theoretical reasons [Ross et al., 2013]. Instead Laura Ross and colleagues showed that an ecological explanation for sex-ratio variation, based on ancestral specialization in roles such as brood rearing, is consistent with the empirical data, concluding that haplodiploidy is not the primary explanation for the origin of eusociality, but rather has played an important role in shaping conflicts within eusocial species [Ross et al., 2013].

The haplodiploidy hypothesis has subsequently been presented as being synonymous with inclusive fitness theory (e.g., [Nowak et al., 2010]), leading some to conclude that disproof of the hypothesis constituted disproof of the underlying theory. In fact, the haplodiploidy hypothesis arose from inclusive fitness theory but is not an integral part of it, and Hamilton himself recognized the evidence accumulating against it.[8] Although the classical haplodiploidy hypothesis[9]

Figure 10.4: Ancestral monogamy predicts the incidence of eusociality among the Hymenoptera. The phylogeny shown is for eusocial hymenopteran genera for which female mating-frequency data are available. Each independent origin of eusociality is indicated by alternate shading. High polyandry (more than 2 effective mates) is denoted by solid gray branches, facultative low polyandry (between 1 and 2 effective mates) by dashed gray branches, and monandry by solid black branches. Monogamy is thus seen to be the basal state in the origin of eusociality, with polyandry appearing later in evolutionary time. Figure redrawn from [Hughes et al., 2008].

ultimately proved to be incorrect, a more modern and more general explanation of the transition to social breeding, the monogamy hypothesis, has proved very fruitful. Like the haplodiploidy hypothesis, this hypothesis is also based on inclusive fitness reasoning, as described below.

10.2.4.2   The Monogamy Hypothesis

In proposing the haplodiploidy hypothesis, Hamilton's starting point was to consider a singly mated queen, since this maximizes relatedness between her female offspring. The monogamy hypothesis, proposed by Jacobus Boomsma [Boomsma, 2007, Boomsma, 2009, Boomsma, 2013], starts from the same observation that relatedness between offspring is maximized when parents mate once and for life. Unlike the haplodiploidy hypothesis, however, the monogamy hypothesis notes that, independent of genetics,[10] under this scenario offspring are equally related to both their siblings and their own offspring.[11] When this is the case, a small increase in the success of raising siblings, relative to the success of raising direct offspring, tips the balance in favor of staying at the home nest to help raise siblings, rather than attempting to reproduce independently. Boomsma thus identifies a "monogamy window" through which species must pass in the evolution of eusociality [Boomsma, 2007, Boomsma, 2009]. Subsequent modifications to the life history of a eusocial species may take place, such as moving towards multiple matings, as occurs in many existing eusocial species; these take place after workers have lost the capacity to mate, however, and all existing eusocial species are predicted to have been monogamous earlier in their evolutionary history.[12]

Empirical support for the monogamy hypothesis is now extensive.[13] Comparative analysis by William Hughes and colleagues has shown how the hypothesis explains the distribution of eusociality among the Hymenoptera (figure 10.4; [Hughes et al., 2008]). For cooperative breeding, analyses by Charlie Cornwallis, Ashleigh Griffin, and colleagues have explained both its evolution and subsequent losses[14] among bird species [Cornwallis et al., 2010], while Dieter Lukas and Tim Clutton-Brock have done the same in mammals [Lukas and Clutton-Brock, 2012]. Roberta Fisher, Charlie Cornwallis, and Stu West have also adapted the monogamy hypothesis to explain obligate and facultative multicellularity,[15] number of cell types, and cell sterility in species representing the archaea, bacteria, and eukaryotes [Fisher et al., 2013].

10.3 SOME FURTHER TOPICS IN SOCIAL EVOLUTION THEORY

Throughout this book the focus has been on relatively simple models of general social behavior, where relatedness between individuals, and effects on fitnesses of individuals and their social partners, can be easily calculated. This abstract approach has helped us to focus on the conceptual foundations of inclusive fitness theory. In the following we briefly discuss how inclusive fitness theory has been applied to understand concrete biological problems.

10.3.1 Sex Allocation

As we have already seen suggestions of above, sex allocation theory has its roots in inclusive fitness theory. Although Fisher initially showed how equal sex ratios could evolve through competition for mates [Fisher, 1930], Hamilton was the first to consider theoretical explanations for observed deviations from equal sex ratio [Hamilton, 1967]. Hamilton considered factors such as inbreeding and haplodiploid genetic systems. As discussed above, shortly afterwards Trivers and Hare proposed that haplodiploidy should generate a conflict between queens and their daughters over the ideal colony sex ratio, with queens preferring equal sex allocation but daughters preferring female-biased sex ratios [Trivers and Hare, 1976]. The subsequent explosion of theoretical and empirical interest in the subject has resulted in a vast literature. Good starting points are Steve Frank's book, which presents three theoretical approaches to sex allocation including the inclusive fitness approach [Frank, 1998], and Stuart West's book which summarizes the theory and the empirical data [West, 2009].

10.3.2 Genetic Kin Recognition

Although genome-wide relatedness arising from population structuring is a natural route towards the evolution of social behavior, mechanisms for directing altruism towards bearers of the same genes can also spread due to inclusive fitness benefits, as in the case of greenbeard traits discussed in chapter 6. Greenbeards are pleiotropic traits simultaneously coding for a marker phenotype, and conditional social behavior towards bearers of the marker. In situations of variable relatedness, and in the absence of such pleiotropy, selective pressure can exist for individuals to make use of existing cues regarding genetic similarity as a trigger for social behavior. However, as

Ross Crozier noted, selection should also erode the genetic diversity underlying such cues, since any individual with the relevant trait will benefit from the social behavior of others, leading to its fixation in the population [Crozier, 1986]. *Crozier's paradox* is thus that markers that start off as good indicators of relatedness become, through the action of selection, poor indicators of relatedness. The paradox was ultimately resolvable only through the construction of sophisticated population genetical models by François Rousset and Denis Roze [Rousset and Roze, 2007], illustrating the need for such approaches in tackling complicated problems of social evolution theory [Gardner and West, 2007].

10.3.3 Spite

In this book we have typically referred to altruistic social behaviors that result in a fitness increment for the target of the behavior. Yet Hamilton also noted the potential for social behaviors to reduce the fitness of others, referring to these as *spite* ([Hamilton, 1964a]; table 2.3). Although Hamilton initially thought that spite would not evolve [Hamilton, 1964a], he subsequently observed that spite can evolve when there is negative relatedness between individuals [Hamilton, 1970].[16] Real biological instances of spite have subsequently been discovered, such as bacterial production of other-harming bacteriocins (figure 1.6). Negative relatedness has a natural interpretation under the geometric view of relatedness (HR4 in chapter 7 and figure 7.2), as arising when interacting with individuals that are less genetically similar to an actor than are random members of the population. The geometric view also enables an interesting change in perspective to be made; spite against negatively related individuals can also be interpreted as altruism towards positively related individuals.[17] Bacteriocins can thus be understood as greenbeard traits that benefit genetic relatives by reducing the competition they experience, by harming less-related individuals [West and Gardner, 2010]. It remains crucially important, however not to confuse behaviors that are purely selfish, and can therefore evolve without any relatedness, with spiteful behaviors requiring negative relatedness [West and Gardner, 2010]; in particular, apparently spiteful behavior can have delayed fitness benefits and therefore actually be selfish [Hamilton, 1970, Foster et al., 2001].

10.3.4 The Evolution of Organismality

The problem of how evolution associates low-level entities into new higher-level entities, that are more than the sum of their parts, is a fundamental one in evolutionary biology. The evidence that this has repeatedly happened is all around us in the natural world, from associations of genes into chromosomes, associations of chromosomes into cells, associations of cells to form endosymbioses such as in the eukaryotic cell, associations of cells to form multicellular organisms, and associations of organisms to form *superorganisms* such as social insect colonies [Bourke, 2011b]. The problem of how such *major evolutionary transitions* occur has been influentially framed by Leo Buss [Buss, 1987], John Maynard Smith and Eörs Szathmáry [Maynard Smith and Szathmáry, 1997], and Richard Michod [Michod, 2000]. Inclusive fitness theory is increasingly important in understanding just how such transitions can arise, as Andrew Bourke's summary of the theory and empirical evidence shows [Bourke, 2011a]. However, inclusive fitness theory has also helped sharpen the often under-appreciated question of what exactly defines an organism; the answer, as proposed by David Queller and Joan Strassmann, is that organisms are defined by a high level of cooperation between their constituent units, and a low level of conflict [Queller and Strassmann, 2009]. Importantly, this means that *organismality* is not a binary property but rather a continuum in two dimensions (figure 10.5).

10.4 OTHER THEORETICAL APPROACHES

This book has focussed on two particular approaches to studying social evolution. The replicator dynamics were introduced in chapter 2 as a way to understand selective pressures and possible evolutionary outcomes in a variety of social scenarios, under the simplifying assumption that individuals have a simple genetically determined social behavior, and that this breeds true, in that offspring always inherit their asexual parent's behavior unchanged. In chapter 3, the Price equation and quantitative genetics were introduced. These tools allow us to consider more complex social behaviors that relate to underlying genes in some unknown way; by making minimal assumptions the instantaneous evolutionary change can be deduced from the selection differential for a trait, and the heritability of that trait by offspring. The Price equation affords us much greater generality, in dealing with any kind of social trait under any kind of genetic control, but apparently at the expense of having a dynamic model of evolution.[18] In particular, this approach allows us to reason, in a simple way, about the fundamental question of evolutionary biology: *why* do traits evolve? To analyze particular traits, however, the Price-equation approach may prove unwieldy. In the following, an undeservedly brief tour of more

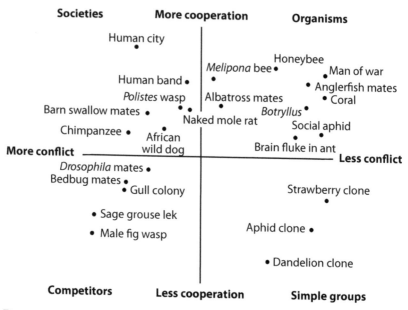

Figure 10.5: David Queller and Joan Strassman propose that "organismality" of groups is a continuous property with two axes: the level of cooperation within the group, and the level of conflict within the group. Organismality is plotted above for various collections of multicellular individuals; under this approach honeybee colonies are organism-like, but naked mole rat colonies much less so, despite both being considered eusocial under some classifications as discussed in chapter 1. Similar classifications can be made for organisms as collections of cells, and for associations between different species. Figure redrawn from [Queller and Strassmann, 2009].

complicated mathematical approaches to studying selection is given, as signposts to the interested reader.

10.4.1 Population Genetics

Population genetics is the study of the dynamics of evolutionary change in populations, taking account of the genetic complexities underlying traits. This means that models need to take account of the particular details of genetic architecture, unlike application of the Price equation, but, also unlike the Price equation, this enables long-term evolutionary dynamics to be predicted.[19] If the true genetic basis of a trait is known (and a proponent of the phenotypic gambit, or a quantitative geneticist, would argue that it can never be truly known for a complex trait), then a population genetical model can capture details that will lead to important differences in evolutionary outcome.[20] Given their dynamic nature, population genetical models can also be used for calculations such as the probability of fixation of an allele, and be used to study nonselective evolutionary phenomena such as genetic drift. The literature is huge, but a classic reference for population genetical models in social evolution is François Rousset's book [Rousset, 2004].[21]

10.4.2 Class-Structured Populations

Throughout this book we have almost exclusively assumed that the individuals in our population all belong to one class, all able to express and receive social behaviors, and all having the same reproductive potential. Only in chapters 8 and 9 have we considered exceptions to this. Yet real biological populations are most likely not to satisfy this assumption, but rather to be *class structured*. In real populations, individuals are likely to vary both in the opportunity they have for expressing social behaviors (chapter 8), and in their potential reproductive value (chapter 9). Clearly for some behaviors these complications need to be incorporated into inclusive fitness models in order to gain appropriate insight. Peter Taylor and Steve Frank presented a general methodology to do this, based on the maximization approach (see below) and neighbor-modulated fitness [Taylor and Frank, 1996, Frank, 1998, Frank, 2013]. With such tools, and some simplifying assumptions (described below), one can tackle more complicated and realistic problems of social evolution, such as sex allocation (discussed above).

10.4.3 Maximization Approaches

As discussed in chapter 2, in evolutionary game theory, populations can be analyzed in terms of evolutionary stable strategies that are local fitness maxima. Peter Taylor and Steve Frank, in developing their class-structured approach to building inclusive fitness models (see above), also adapted the *maximization approach* inherent in evolutionary stability analysis for use in their models; by treating fitness costs, benefits, and associations between like individuals all as derivatives, Taylor and Frank were able to apply standard calculus to derive equilibrium points under selection, and explain these in inclusive fitness terms [Taylor and Frank, 1996, Frank, 1998, Frank, 2013]. Moving between regression coefficients and derivatives requires an assumption, however: there must be small variances in the predictors involved.[22] This translates into an assumption of weak selection, since if there were significant variance in genetic values in a population this would normally also result in significant variance in fitnesses. This assumption, required by a popular and effective approach to building inclusive fitness models, has caused some confusion however, as discussed below.

10.4.4 Concepts versus Tools

The focus of this book, as mentioned throughout, has been on the core concepts of inclusive fitness theory. Above we have seen a small number of particular mathematical tools that can be used to analyze particular social evolution scenarios. Some of the recent criticisms of inclusive fitness theory appear to have conflated conceptual issues with methodological issues.[23] Thus, the claim that inclusive fitness theory requires weak selection [Nowak et al., 2010], for example, may arise from conflating the relatedness in Hamilton's rule with pedigree relatedness (see chapter 4), or it may also arise from the assumptions used in approaches such as that of Peter Taylor and Steve Frank (see above) [Taylor and Frank, 1996]. It is therefore crucial to avoid confusion between conceptual models of social evolution, and the methodological approaches to analyzing particular problems of social evolution [Gardner et al., 2011, Marshall, 2011a]. Inclusive fitness is the fundamental concept, while the mathematical tools we apply to reason about it, such as the Price equation or the replicator dynamics, are just that.

10.5 CONCLUSION

Fifty years after its introduction, inclusive fitness theory retains its status as the general theory of biological evolution that Hamilton felt it would prove to be.[24] Empirically, inclusive fitness theory is probably one of the most tested, and supported, modern biological theories (e.g., [Abbot et al., 2011, Bourke, 2011b, Bourke, 2014]). Theoretically, inclusive fitness is, as Hamilton thought, the most important extension of evolutionary theory since Fisher. As this book has shown, the theory is general, in that it can be applied to explain the evolution of arbitrary social scenarios, and in that it also contains the classical evolutionary theory of asocial traits as a special case. The scenarios considered in this book have been simple, to illustrate the fundamental concepts involved. When applied to more complicated evolutionary scenarios, such as the examples given immediately above, analyses of the kind presented here may prove impractical. However, having established the fundamentals in the simplest cases, we should have confidence that inclusive fitness also governs evolution in more complex scenarios. Such an approach is not peculiar to the mathematical description of evolutionary scenarios. In domains from applied mechanics through to physics, it is commonplace to analyze systems where the equations governing behavior are well known and apparently simple, yet to find that the mathematical tools we can successfully apply in the analysis rest on approximations or the solution of vastly simplified problems. No researcher in dynamical systems or theoretical physics would suggest this should lead us to question the correctness of the equations we started with; the limitation is rather with the tools we use to investigate them. The study of social evolution theory is no different.

GLOSSARY

additive fitness effects. Fitness effects that interact additively, in that the net fitness effect on an individual is simply the sum of the component fitness effects, such as those due to the behavior of that individual and of their social partners (see also **nonadditive fitness effects**).

altruism. A behavior or trait that, on average, is costly in terms of lifetime **direct fitness** to express or bear (see also **mutation test**).

breeding value. A linear combination of **genic values** for a particular quantitative trait, corresponding to the average offspring value of that trait when mated randomly into a reference population.

classical fitness. The original conception of fitness in terms of direct reproduction of individuals (**direct fitness**), implicitly excluding fitness effects due to social partners. Not equivalent to **neighbor-modulated fitness**.

coefficient of relatedness. The correct measure of **relatedness** for **inclusive fitness** theory, as a regression coefficient predicting **genetic value** of social partners according to **genetic value** of a focal individual.

coefficient of relationship. The correlation coefficient between **genetic values** of a focal individual and its social partners, sometimes confused with the correct measure of **relatedness** for **inclusive fitness** theory, the **coefficient of relatedness**.

compensated relatedness. A modified version of the **coefficient of relatedness** in inclusive fitness theory, in which **relatedness** to those displaced by the offspring of recipients of a social behavior is subtracted from the **relatedness** to those recipients.

cooperation. Social behavior providing a **direct fitness** benefit to the performing individual but also a, possibly larger, fitness benefit to social partners (see also **mutual benefit**). Used by some other authors to refer to both **altruism** and **mutual benefit**.

direct fitness. The portion of an individual's **inclusive fitness** due to their own direct reproduction as caused by their traits or behavior, excluding any fitness effects attributable to social partners.

donation game. The simplest game in which an individual can sacrifice **direct fitness** in order to increase the fitness of a single social partner. Can be formulated in both **additive** and **nonadditive** versions, and by selecting appropriate parameter values can describe **altruism**, **cooperation**, **spite**, or **selfishness**.

falsebeard. A trait coding for a conspicuous **phenotypic** marker that triggers **donation** by **greenbeards**, without coding for **donation** towards bearers of that **phenotype**. Where this dissociation between marker and **donation** is possible, **greenbeards** are susceptible to replacement by falsebeards.

135

genetic kin recognition. The use of a phenotypic marker to infer common genetic ancestry, and hence increased genome-wide genetic **relatedness**.

genetic value. A linear combination of **genic values**, such as **breeding value**; a single **genic value** is a special case of genetic value.

genic value. A value indicating presence, or dosage, of a particular allele. For a diploid trait, for example, genic value could be a count of the number of copies of an allele (0, 1, or 2).

geometric view of relatedness. A geometric interpretation of the **coefficient of relatedness**, in which **relatedness** is conceptualized as the relative position of a social partner's (or social partners') expected **genetic value** between that of a focal individual and the population mean.

greenbeard. A trait simultaneously coding for a conspicuous **phenotypic** marker, such as a green beard, and conditional **donation** towards bearers of that marker. Unlike **genetic kin recognition**, **greenbeard** mechanisms only detect **relatedness** at a subset of the genome (but see also **falsebeards**). Can be obligate or facultative, helping or harming.

group selection. A view of social evolution as competition between groups, in which selective pressure on a trait has inter- and intragroup components, often having opposite directions. Correct predictions of evolutionary change made using group selection analyses necessarily agree with those made using **inclusive fitness** theory. Also known as multi-level selection.

heritability, narrow-sense. A quantitative measure of the **phenotypic** resemblance between parent and offspring (whether direct or indirect), also defined as the coefficient of the regression of an individual's **genetic value** on their expressed **phenotype** for that trait.

identity by descent. Situation in which copies of alleles within an individual, or within multiple individuals, arise from the same common ancestor. Can be used to estimate **relatedness**, but only in situations of weak selection.

inclusive fitness. An extension of classical fitness to account for social traits and behaviors. Defined as the sum of **direct fitness** (excluding effects due to social partners) and **indirect fitness**.

indirect fitness. The component of an individual's **inclusive fitness** that is due to the increased direct reproduction of social partners, as facilitated by the focal individual, and weighted by the **relatedness** between the individual and the partners.

kin selection. A label for **inclusive fitness** theory, coined by John Maynard Smith, and often used as a synonym for it. Can also mean, more technically, **inclusive-fitness**-based selection within family groups.

mutation test. A test to classify whether a trait or behavior is directly beneficial to the bearer. Can be formulated in genetic (trait) or **phenotypic** (behavior) forms.

mutual benefit. A behavior or trait that is beneficial to the **direct fitness** of the bearer, and also increases the fitness of social partners.

neighbor-modulated fitness. An alternative way of accounting for fitness effects in social evolution, in which the effects of a focal individual on their **direct fitness** as well as the effects of social partners on that individual's fitness are summed. Frequently provides mathematical convenience and when applied appropriately is equivalent to **inclusive fitness**. Not conceptually equivalent to **classical fitness**.

nonadditive fitness effects. Fitness effects that interact nonadditively, in that the net fitness effect is more than, or less than, the simple sum of the component fitness effects. Nonadditive fitness effects can lead to substantially different selective pressures to **additive fitness effects**.

phenotype. The expressed level of a trait or behavior, usually determined at least partially by **genetic value**.

phenotypic gambit. The strategy of modeling a trait of interest as if it had the simplest possible genetic basis, a single haploid locus.

public goods game. A generalization of the **donation game** to groups of more than two individuals. Can also be formulated in versions with **additive** or **nonadditive fitness effects**.

relatedness. The quantity required in **inclusive fitness** theory to determine the relative valuation of an individual's **direct** and **indirect fitness**. Properly calculated as the **coefficient of relatedness** (which has a **geometric** interpretation) or, equivalently, a ratio of **heritabilities**. Can also be estimated from pedigrees under appropriate conditions (see **identity by descent**).

reproductive value. The long-term evolutionary contribution of an individual to future generations; frequently does not correspond with the number of immediate offspring produced.

selfish. A trait or behavior that increases the bearer's **direct fitness** while having an incidental negative (or zero) effect on the fitness of others in the population.

spite. A behavior or trait that, on average, is costly in terms of lifetime **direct fitness** to express or bear, while having a negative effect on the fitness of social partners.

weak altruism. See **mutual benefit**. A misnomer, since "weak altruism" actually has **direct fitness** benefits for the bearer (see **altruism**).

NOTES

CHAPTER 1. SOCIAL BEHAVIOR AND EVOLUTIONARY THOUGHT

1. Darwin employed several collectors to send animal specimens to him, including Wallace; Wallace also was a prodigious collector on his own account, but his collection was destroyed at the conclusion of a five-year voyage when the ship containing it caught fire [Raby, 2002].

2. There is some disagreement, however, over whether existence of castes, for example, are relevant for mammals, and hence whether naked mole rats and several other mammalian species actually qualify as eusocial after all; see, for example, [Burda et al., 2000].

3. Aposematism may be more complex than a purely self-sacrificing trait, since otherwise the initial rarity of aposematic morphs, and the delay in predator learning, would make establishment of the trait unlikely under many scenarios; see, for example, [Speed, 2001].

4. Excepting an earlier book review, as noted by Hamilton [Hamilton, 1996, p. 1].

5. This book primarily celebrates the achievements of Hamilton in formulating inclusive fitness theory, and the generality of his approach. For further reading on "Bill" Hamilton see his collected papers and accompanying essays [Hamilton, 1996, Hamilton, 2001, Hamilton, 2005], as well as his biography by Ullica Segerstrale [Segerstrale, 2013].

6. As discussed in chapter 9 (note 1), Haldane's focus seems to have been not on altruism per se, but rather on the effects of rare genes on evolution. For a detailed discussion of Haldane's intentions, and Hamilton's views on his own priority, see [Segerstrale, 2013].

7. Note also, however, that the evolution of mutually beneficial behaviors (see chapter 2) is also predicted by condition (1.3), so its satisfaction does not guarantee that the behavior experiencing positive selection is the altruism Hamilton initially sought to explain [Foster, 2008].

CHAPTER 2. MODELS OF SOCIAL BEHAVIOR

1. Von Neumann had actually laid the foundations for the theory during the previous decades, beginning in 1928 with the publication of a single paper in German [Von Neumann and Morgenstern, 1944, p. 1].

2. The Nash equilibrium concept has also had an impact on popular culture, spawning the book *A Beautiful Mind* about the life of John Nash [Nasar, 1998], and the film of the same name.

3. The cognitive scientist Herbert Simon proposed that models of choice needed to explain departures from economic rationality observed in real decision makers, introducing the concept of *bounded rationality* [Simon, 1956].

4. "Interesting" in this context means that the game structure is such that the players cannot arrive at the mutually and individually optimal outcome through simple strategies, such as "always play I."

Of the games identified as "interesting" by [Rapoport, 1967], only "prisoner's dilemma" and "chicken" have received significant attention from evolutionary biologists. Note that games with a positive net effect on individual direct fitness from a behavior do not represent interesting games, and hence do not correspond to any economic game in table 2.4.

5. The derivation of equation (2.1) is straightforward given the standard definition of the replicator dynamics,

$$\frac{df}{dt} = f\left(w_{\mathrm{I}} - (fw_{\mathrm{I}} + (1-f)w_{\mathrm{II}})\right). \tag{2.4}$$

The average payoff of an action **I** player in the population is given by

$$w_{\mathrm{I}} = -c + fb, \tag{2.5}$$

since these players always donate and pay the cost c, but meet other action **I** players and receive the benefit b at the population relative frequency f. The average payoff of an action **II** player is the same, except that they never donate and so never pay the cost c, hence

$$w_{\mathrm{II}} = fb. \tag{2.6}$$

Substituting equations (2.5) and (2.6) into equation (2.4) gives equation (2.1) in the main text.

6. There are two important mathematical caveats to this statement: (i) If $f = 1$ then it will not decrease, because when all the population is playing **I** there are no **II** players to reproduce and thus outcompete those **I** players. However, if an infinitesimal number of **II** players arose, through immigration into the population or through mutation, for example, those would take over the population, driving the **I** players to extinction. Thus since the $f = 1$ equilibrium is unstable if perturbed, it does not correspond to an equilibrium we would expect to see a real biological population occupying. (ii) Because of the nature of differential equations, f actually approaches 0 at a slower and slower speed, but never reaches it. This is due to an assumption of the replicator dynamics, that the population in question is infinite. Since real biological populations are always finite, eventually the last **I** player would die and f would reach 0.

7. Expanding out the brackets in equation (2.1), canceling terms, and simplifying shows that

$$\frac{df}{dt} = -f(1-f)c. \tag{2.7}$$

When $0 < f < 1$ and $c > 0$, then equation (2.7) is always negative, regardless of the particular values of these variables. Note that the benefit from donation, b, does not appear in equation (2.7), and hence is irrelevant.

8. The assumption of random interaction in the replicator dynamics manifests in the fact that the expected payoffs of each strategy (equations (2.5) and (2.6), for example) are calculated according to the frequencies with which different pairings arise, calculated simply according to the products of the frequencies of each strategy type in the population; this corresponds to members of the population being paired completely at random irrespective of their strategy.

9. The derivation of equation (2.2) follows that of equation (2.1) as detailed in note 5. This time, the average payoff of an action **I** player in the population is given by

$$w_{\mathrm{I}} = -c + f(b + d), \tag{2.8}$$

since when both players choose action **I** each now receives $b + d$. The payoff of an action **II** player is unchanged from equation (2.6). Substituting equations (2.8) and (2.6) into equation (2.4) gives equation (2.2) in the main text.

10. The equilibria of the replicator dynamics correspond to values of population relative frequency f where the rate of change df/dt is 0. Thus by finding the roots of equation (2.2) one finds the population equilibria under the replicator dynamics. Some algebra shows these to be $f = 0$, $f = 1$ (in fact these first two are *always* equilibria of the replicator dynamics; see also note 6), and $f = c/d$.

11. Some authors refer to synergism as *positive* deviation from additivity ($d > 0$); however here we will follow Queller's original use of the term [Queller, 1984, Queller, 1985].

12. A simple technique to check the stability of a fixed point in our one-dimensional replicator dynamics is to take the differential equation describing the replicator dynamics for our particular payoff matrix, differentiate it again with respect to the frequency f of strategy **I**, and then evaluate the resulting derivative at the fixed point. Formally, for the replicator dynamics of equation (2.2) whose mixed equilibrium we have already calculated as $f = c/d$, we are interested in

$$\left. \frac{d}{df} \right|_{f=c/d} f\left(-c + f(b + d) - (f(-c + f(b + d)) + (1 - f)fb)\right). \tag{2.9}$$

That is, we want to find the sign of expression (2.9). Why do we want to do this? We know that at the fixed point $f = c/d$ there is no change in f due to selection, so $df/dt = 0$. What we want to work out then is, if we move away from the fixed point by varying f, is the action of selection going to push us back to the equilibrium or away from it? We can determine this from the slope of the function df/dt as it goes through $f = c/d$, which is what expression (2.9) gives us; if the slope is positive then when f is increased the replicator dynamics df/dt will increase it further over time, and when f is decreased the replicator dynamics will reduce it further over time. Thus, if the slope of df/dt is positive around $f = c/d$ then the fixed point is unstable. On the other hand, if the slope is negative, then the opposite behavior occurs: the replicator dynamics df/dt tend to return f to c/d when disturbed, and the fixed point is asymptotically stable.

Having explained the procedure, we now work out the answer to our original question. Evaluating expression (2.9) and simplifying gives us

$$c\left(1 - \frac{c}{d}\right), \tag{2.10}$$

which, since $c > 0$, is positive when $c < d$ (since then $1 - c/d$ is positive), telling us that the equilibrium is unstable. As $f = c/d$ corresponds to a real population frequency (between 0 and 1) only when $c < d$, we care about that case only, hence when the equilibrium frequency $f = c/d$ is one that the population can reach, that equilibrium is always unstable.

13. To see that costly donation with diminishing returns is always selected against, we take the replicator dynamics described in equation (2.2), which we can rearrange to give

$$\frac{df}{dt} = f(1 - f)(fd - c). \tag{2.11}$$

Now since f is always between 0 and 1 inclusive, $f(1 - f)$ is always positive, so whether equation (2.11) is positive depends on whether $fd - c$ is positive. Since we are interested in costly donation ($c > 0$) with negative nonadditivity ($d < 0$) this expression is always negative, hence equation (2.11) is always negative and costly donation with diminishing returns always experiences negative selection.

14. Behavioral effects on fitness could also be neutral. Very few effects are likely to be completely neutral, however; even behaviors that might appear to be asocial, such as foraging, are likely to affect the fitness of others in some way, through reducing food available for conspecifics, for example. Therefore, neutral effects are not considered in the classification of table 2.3.

15. Mutually beneficial behavior has historically been referred to as *weak altruism* [Wilson, 1979], and also as *by-product mutualism* or cooperation [West et al., 2007b]. In this book we will follow the practice of others in referring to such behavior wherever possible as behavior that produces mutual benefit, since reference to weak altruism can lead to confusion with strong altruism, whereas the two actually require different classes of evolutionary explanation [West et al., 2007b]. Similarly, as [West et al., 2007b] does, we will avoid by-product mutualism in order to distinguish cooperative behavior, which occurs within a single species, from mutualism occurring between species, such as in the association between plants and their pollinators. In some cases however "cooperation" is a convenient shorthand, intended in this book to mean the same as "behavior producing mutual benefit."

16. That cooperation in unstructured populations experiences positive selection, even if cooperation increases the fitness of others more than it increases the fitness of behaving individuals, is seen from
equation (2.7). This equation is positive (so cooperation experiences positive selection) whenever "cost"
of cooperation c is positive, while benefit to others b does not appear and is therefore irrelevant.

17. The existence of a stable mixed equilibrium for cooperation with diminishing returns is
established as follows. First, analysis of equation (2.2) has already shown us that there is a population
equilibrium when $f = c/d$. Since for cooperation $c < 0$ then, for this equilibrium to be one the
population can reach (i.e., between 0 and 1), deviation from additivity d must be negative and smaller
than c (i.e., $d < c < 0$). We next check the stability of the equilibrium, as described above in note 12.
By looking at expression (2.9) we see that it is negative when c is negative, provided that $c/d < 1$. As
argued above, when expression (2.9) is negative the corresponding equilibrium is stable, so cooperation
with diminishing returns can have a stable mixed equilibrium.

18. Hauert et al.'s formulation of nonadditive public goods games continues a tradition started by Uzi Motro [Motro, 1991]. The historical roots of such models are reviewed in
[Archetti and Scheuring, 2012].

19. In Hauert et al.'s game the public good in the group is divided equally between all members, and
each successive contribution to the public good b is multiplied by a constant δ, so the second contribution
to the group is multiplied by δ^2, the third by δ^3, and so on. The constant δ thus controls the additivity of
the game with (assuming that the individuals contribute positively to the public good, so $b > 0$) $\delta < 1$
capturing negative nonadditivity (diminishing returns), $\delta = 1$ capturing additivity, and $\delta > 1$ capturing
positive nonadditivity. The following expression defines the payoff to a group member in a group with n
members of whom k are donators:

$$\frac{b(1 - \delta^k)}{n(1 - \delta)}. \tag{2.12}$$

Hauert et al. present a model in which groups of size n are formed at random from the population
[Hauert et al., 2006], and using expression (2.12) calculate the expected fitness of social (type **I**) and
asocial (type **II**) individuals as

$$w_{\mathbf{I}} = \frac{b}{n(1 - \delta)} \left(1 - \delta(1 - f + \delta f)^{n-1}\right) - c \tag{2.13}$$

and

$$w_{\mathbf{II}} = \frac{b}{n(1 - \delta)} \left(1 - (1 - f + \delta f)^{n-1}\right), \tag{2.14}$$

respectively. One can find the fixed points of the replicator dynamics for equations (2.13) and (2.14)
by simply solving for when they are equal. Using this approach Hauert et al. show [Hauert et al., 2006]
that if the net effect of contributing to the public good on an individual's own direct fitness, excluding
potential nonadditivity, is negative (i.e., if $b/n - c < 0$) then asocial types go to fixation, unless there
is large enough positive nonadditivity ($\delta^{n-1} > cn/b$), in which case either social or asocial types go to
fixation in the population according to their initial frequency. Similarly, if an individual's contribution to
the group, excluding any nonadditivity, has the net effect of increasing its direct fitness (i.e., if $b/n - c >$
0) then selection favors social types, unless there is sufficient negative nonadditivity ($\delta^{n-1} < cn/b$),
in which case a mixed-population equilibrium containing both social and asocial types exists. These
different evolutionary outcomes, and their dependence on the net effect of an individual's behavior on
its own net direct fitness, and the deviation from additivity of interactions, are exactly those shown for
two-player interactions and summarized in table 2.4.

One last point: the observant reader may have noted that the public goods payoff function (2.12)
appears to be undefined in the additive case, when $\delta = 1$, since then the denominator of (2.12) is 0. To
correctly work out the additive payoff function from (2.12) we apply L'Hôpital's rule, which states that
we can work out the value of (2.12) as linearity ($\delta = 1$) is approached by differentiating the numerator

and denominator of the fraction with respect to δ, and then evaluating the resulting fraction with $\delta = 1$. This gives us

$$\frac{-bk\delta^{k-1}}{-n},\tag{2.15}$$

which, when $\delta = 1$, gives us the linear public goods game payoff function

$$\frac{bk}{n}.\tag{2.16}$$

20. The results of [Hauert et al., 2006] for behaviors that make a positive contribution to the public good are described in note 9. To determine how, for behaviors that reduce group welfare such as spite and selfishness, the evolutionary outcomes depend on the net effects of behavior on a focal individual's direct fitness, and on the deviation from additivity of interactions, we must repeat that analysis but under the assumption that $b < 0$.

To find when there is a stable mixed equilibrium of social and asocial types, or *bistability* such that populations containing all social types or all asocial types are both stable, we solve for the value of f at which equations (2.13) and (2.14) are equal, find conditions when this equilibrium value of f lies between 0 and 1, then check the stability of that equilibrium point. When the fitnesses of the two types, social and asocial, are equal then we necessarily have a fixed point of the replicator dynamics. Algebra shows that when $\delta > 1$ the mixed-population equilibrium exists when

$$1 < \frac{cn}{b} < \delta^{n-1},\tag{2.17}$$

while for $\delta < 1$ it exists when

$$\delta^{n-1} < \frac{cn}{b} < 1.\tag{2.18}$$

We can check the stability of the equilibrium as in note 12 but, since we are not explicitly using the replicator dynamics, we instead differentiate the difference between the fitness of social types (2.13) and asocial types (2.14) with respect to f, evaluated at the equilibrium f^*. That way, if f increases above f^* and social types increase in fitness as a result, then further increases in f will further increase this fitness difference and lead to fixation of the social type under the replicator dynamics, whereas if f reduces from f^* then the social type will become less fit that the asocial type, and the asocial type will go to fixation. Thus, if the derivative we calculate and evaluate at the equilibrium is positive then the equilibrium is unstable, while if the derivative is negative then the equilibrium is stable, just as in note 12.

Differentiating $((2.13) - (2.14))$ and evaluating at the equilibrium shows that the equilibrium is stable if $\delta > 1$ and unstable if $\delta < 1$. From inequalities (2.17) and (2.18) we can, for negative benefit $(b < 0)$, derive the same results presented in [Hauert et al., 2006] for positive benefit and summarized in note 19. However, it seems as if the stabilities of the equilibria are opposite. For positive benefits, Hauert et al. show that equilibria can be stable only if $\delta < 1$ (negative nonadditivity) and can be unstable only if $\delta > 1$ (positive nonadditivity), whereas when benefits are negative then only $\delta > 1$ allows stable mixed equilibria, and only $\delta < 1$ allows unstable mixed equilibria. Yet, it must be remembered that when benefits are negative, the effect of δ is reversed, so $\delta > 1$ actually implements *negative* nonadditivity, and $\delta < 1$ actually implements *positive nonadditivity*. Thus the possible evolutionary outcomes when public "benefits" are negative occur according to precisely the criteria that predict them in positive-benefits public goods games, and in pairwise interactions, as summarized in table 2.4.

21. Archetti and Scheuring's group payoff is defined as

$$\frac{1}{1 + e^{-s(k-m)}},\tag{2.19}$$

where k is the number of individuals contributing to the group benefit, and m is the threshold required to produce the group benefit; all individuals receive the same group benefit, but those contributing to it also pay a cost c [Archetti and Scheuring, 2011]. The parameter s controls the nature of the dilemma, interpolating the sharp threshold of the original formulation in the social sciences [Dieckmann, 1985] ($s = \infty$), and the pure additive public goods game (s approaches 0, but note that at $s = 0$ there is no dependence of public good on the number of volunteers), as illustrated in figure 2.6.

22. The approximation in equation (2.3) is for large group sizes [Archetti and Scheuring, 2011].

CHAPTER 3. THE PRICE EQUATION

1. George R. Price was one of the most intriguing figures of twentieth-century evolutionary biology. Various authors have written on his life and achievements [Frank, 1995, Schwartz, 2000, Gardner, 2008, Harman, 2010].

2. Price took care to note that he was unaware of any previous formulation of selection using his approach. In fact, Alan Robertson in 1966, and Ching Chun Li in 1967, had already presented similar results. Robertson's paper [Robertson, 1966] was in a somewhat obscure venue, but Li's [Li, 1967] was in *Nature*, as was Price's [Price, 1970]. Both Robertson and Li, however, both presented only the covariance/regression part of the Price equation, thereby providing an equation that is only correct under certain conditions as we shall see later, and not applicable to an important range of problems such as those involving multilevel selection [Price, 1972a]. Robertson considered his theorem important enough to call it the "secondary theorem of natural selection" [Robertson, 1968], although arguably it is more important than Fisher's "fundamental theorem."

3. The genic value is that of a particular gene on the genotype, as distinct from the value of the entire genotype.

4. In fact, change in mean genotypic value could also be studied using the Price equation, through assigning a unique and arbitrary value to each possible genotypic combination of alleles. This has less utility in evolutionary reasoning than studying additive functions of genic values, however.

5. The derivation of the Price equation proceeds as follows. We wish to calculate

$$\Delta g = \overline{g'} - \overline{g} = \frac{1}{n} \sum_{i=1}^{n} \frac{w_i}{\overline{w}} g_i' - \frac{1}{n} \sum_{i=1}^{n} g_i, \qquad (3.10)$$

which we multiply by $\overline{w}$ to give

$$\overline{w} \Delta g = \frac{1}{n} \sum_{i=1}^{n} w_i g_i' - \frac{1}{n} \sum_{i=1}^{n} \overline{w} g_i. \qquad (3.11)$$

Since the average offspring genic value for the ith individual equals the parent's genic value plus the deviation between the parent's value and the average offspring value, so $g_i' = g_i + \Delta g_i$, we can rewrite equation (3.11) as

$$\overline{w} \Delta g = \frac{1}{n} \sum_{i=1}^{n} w_i (g_i + \Delta g_i) - \frac{1}{n} \sum_{i=1}^{n} \overline{w} g_i, \qquad (3.12)$$

which can be rearranged to give

$$\overline{w} \Delta g = \frac{1}{n} \sum_{i=1}^{n} g_i (w_i - \overline{w}) + \frac{1}{n} \sum_{i=1}^{n} w_i \Delta g_i. \qquad (3.13)$$

The second term of equation (3.13) is the average change in the offspring genic values, weighted by offspring frequency, so is equal to the evolutionary change due to transmission, $\overline{\Delta g}$, in the Price equation (3.1). The first term is the statistical covariance between the g and w pairs, so is the $\text{Cov}(g, w)$ term in the Price equation (3.1). The first term may not look like the more familiar definition of covariance (see note 7), although it is generally used when working with the Price equation, and can be shown to be equivalent as follows. First we take the standard statistical definition of covariance, as in equation (3.19), and expand it out as

$$\text{Cov}(g, w) = \frac{1}{n}\left(\sum_{i|g_i=0}(-\overline{g}(w_i - \overline{w})) + \sum_{j|g_j \neq 0}(g_j(w_j - \overline{w}) - \overline{g}(w_j - \overline{w}))\right), \tag{3.14}$$

so we have divided the population of n parents into those whose genic value is zero ($g_i = 0$) and those whose genic value is nonzero ($g_j \neq 0$). We rearrange equation (3.14) as

$$\text{Cov}(g, w) = \frac{1}{n}\left(\sum_{i=1}^{n}(-\overline{g}(w_i - \overline{w})) + \sum_{j|g_j\neq 0} g_j(w_j - \overline{w})\right). \tag{3.15}$$

We can then rewrite the first term in equation (3.15) as

$$\frac{1}{n}\sum_{i=1}^{n}(-\overline{g}(w_i - \overline{w})) = -\overline{g}\left(\frac{1}{n}\sum_{i=1}^{n}w_i - \frac{n\overline{w}}{n}\right)$$
$$= -\overline{g}(\overline{w} - \overline{w}) \tag{3.16}$$
$$= 0.$$

Also, since $g_j(w_j - \overline{w}) = 0$ when $g_j = 0$, we can rewrite the second term as $\sum_{j=1}^{n} g_j(w_j - \overline{w})$. Thus we have shown that

$$\text{Cov}(g, w) = \frac{1}{n}\sum_{i=1}^{n}g_i(w_i - \overline{w}), \tag{3.17}$$

which completes our derivation of the Price equation (3.1).

6. Thus, if there are n parents in the parental population then

$$\overline{w} = \frac{1}{n}\sum_{i=1}^{n}w_i. \tag{3.18}$$

The average parental genic value $\overline{g}$ is defined similarly, but the average offspring genic value $\overline{g'}$, and the average difference between parental and average offspring genic value $\overline{\Delta g}$, are actually weighted averages (see note 8).

7. The standard statistical definition of the estimate of the covariance between parents' genic values g_i and offspring numbers w_i is, assuming n individuals in the parental population,

$$\text{Cov}(g, w) = \frac{1}{n-1}\sum_{i=1}^{n}(g_i - \overline{g})(w_i - \overline{w}). \tag{3.19}$$

This is not quite the covariance that appears in the Price equation, where the coefficient is $1/n$ rather than $1/(n-1)$ [van Veelen, 2005]. Price, however, was concerned with the theoretical application of the Price-equation approach in models, i.e., to statistical populations [Price, 1972a], as pointed out by Steve Frank [Frank, 2012]; as sample sizes approach the infinite population, $1/(n-1)$ converges to $1/n$. Price made clear that he was considering statistical populations, and therefore chose not to distinguish sample and population (random variable) notation [Price, 1972a], as has been done in this chapter. When the Price equation is applied to experimental data it is applied as an estimator, and many organismal attributes, such as breeding value, must also be estimated; this raises a number of important statistical considerations (e.g., [Hadfield et al., 2010, Morrissey et al., 2010]) which are beyond the scope of this book's theoretical focus.

8. The weighted averages relating to the offspring population are thus

$$\overline{g'} = \frac{1}{n} \sum_{i=1}^{n} \frac{w_i}{\overline{w}} g_i' \tag{3.20}$$

and

$$\overline{\Delta g} = \frac{1}{n} \sum_{i=1}^{n} w_i \Delta g_i. \tag{3.21}$$

These are sometimes referred to as fitness-weighted averages, although fecundity-weighted averages would be more accurate as the weighting is done by immediate offspring numbers, which need not coincide with the proper definition of evolutionary fitness, as we shall see in chapter 10.

9. So $\Delta g = \overline{g'} - \overline{g}$.

10. Some do question whether the Price equation really does achieve the partitioning into change due to selection and change due to transmission bias claimed in the main text. See, for example, [Okasha, 2006] and [Birch, 2013]; the philosophical nuances of this debate are beyond the scope of the present book, however.

11. That is, apart from random fluctuations around 0 due to small sample sizes.

12. The coefficient of the linear regression of y on x, computed using the least-squares method (e.g., [Grafen and Hails, 2002]), is the covariance of x and y, divided by the variance in x; that is,

$$\beta_{yx} = \frac{\text{Cov}(x, y)}{\text{Var}(x)}. \tag{3.22}$$

Hence the Price equation (3.1) can be written in a regression coefficient form [Price, 1970]:

$$\overline{w} \Delta g = \beta_{wg} \text{Var}(g) + \overline{\Delta g}. \tag{3.23}$$

Since variances are never negative, the direction and strength of selection is proportional to the slope of a linear regression of offspring quantity on parental genic value, as discussed in the main text.

13. Figure 3.1 does not quite pertain to Price's hypothetical example of why deer shed and renew their antlers, but it is satisfyingly close.

14. There are frequently important technical details and obstacles in directly determining individuals' genetic composition for a trait, however. For example, see [Falconer and Mackay, 1996, chapter 21].

15. Or, equivalently, whether the slope of a linear regression of offspring production on genic value (or a function thereof) is positive or negative; see note 12.

16. Here there is an unfortunate potential for terminological confusion, since a *statistical population* is not the same as a *biological population* on which we might compute some statistics; a biological population would actually correspond to a *statistical sample* from the *statistical population*.

17. Expectations and variances of discrete random variables, as studied in this book, are very similar to sample averages and variances. To take the example of the coin-toss random variable C, which has a probability p of coming up heads ($C = 1$), and probability $(1 - p)$ of coming up tails ($C = 0$), the expectation of C is calculated as

$$E(C) = 0 \times P(C = 0) + 1 \times P(C = 1) = 0 \times (1 - p) + 1 \times p = p, \tag{3.24}$$

where, for example, $P(C = 1)$ is the probability that the coin toss result is heads.

There is a simple route to calculating (co)variances for random variables, using only expectations. If we have two random variables X and Y then

$$\text{Cov}(X, Y) = E(XY) - E(X)E(Y), \tag{3.25}$$

where $E(XY)$ is the *joint expectation* of X and Y, and $E(X)$ and $E(Y)$ are the *marginal expectations* of X and Y, respectively; there is an equivalent version of this equation for the covariance of two random samples. Since the variance of a random variable X is its covariance with itself (and similarly for the variance of a sample), we can calculate the variance of our coin toss random variable C as

$$\text{Var}(C) = \text{Cov}(C, C) = \text{E}(C^2) - \text{E}(C)^2 = p - p^2 = p(1 - p), \tag{3.26}$$

since $1^2 = 1$.

18. Note that in equation (3.3) the change in the expected value of the population's genetic value G in response to selection is computed ($\Delta\text{E}(G)$). All other expectations and covariances in this version of the Price equation, however, refer to population values for genetic values and offspring numbers in the present generation. Since the former ($\Delta\text{E}(G)$) does not appear in applications of the Price equation to particular scenarios, which is the main focus of this book, this distinction is not highlighted notationally.

19. C. C. Li was among the first to attempt to generalize Fisher's theorem [Li, 1967]. The history of the theorem's misinterpretation and subsequent derivation by G. R. Price is reviewed in [Frank and Slatkin, 1992, Frank, 1995].

20. Evolutionary change in both fitness due to selection, and fitness due to environmental change, are defined respectively as $\Delta_S\text{E}(W) = \text{E}(W'|E) - \text{E}(W|E)$ and $\Delta_E\text{E}(W) = \text{E}(W'|E') - \text{E}(W'|E)$, where W' and W refer respectively to offspring generation fitness and parent generation fitness, as usual, and E' and E refer to the environment experienced by the offspring and the parent generations, respectively. Thus $\text{E}(W'|E)$ is the *conditional expected* fitness of the offspring generation, if it were to experience the parental environment, and so on. Adding these two evolutionary changes together, as in equation (3.5), gives us the total evolutionary change due to selection and due to environmental change that we would intuitively expect, since

$$\Delta\text{E}(W) = \Delta_S\text{E}(W) + \Delta_E\text{E}(W)$$

$$= \text{E}(W'|E) - \text{E}(W|E) + \text{E}(W'|E') - \text{E}(W'|E) \tag{3.27}$$

$$= \text{E}(W'|E') - \text{E}(W|E),$$

so total evolutionary change is the difference between the expected fitness of the offspring generation given they experience the offspring environment, and the expected fitness of the parent generation given they experienced the parental environment.

21. Equation (3.6) is derived as follows. First, we concentrate only on the equation for change in fitness due to selection $\Delta_S\text{E}(W)$ which is a version of the Price equation (3.3), hence

$$\Delta_S\text{E}(W) = \text{E}(W'|E) - \text{E}(W|E)$$

$$= \frac{1}{\text{E}(W)}\left(\text{Cov}(W|E, W) + \text{E}(W\Delta W|E)\right) \tag{3.28}$$

$$= \text{Cov}\left(\frac{W|E}{\text{E}(W)}, \frac{W}{\text{E}(W)}\right) + \text{E}\left(\frac{W}{\text{E}(W)}\Delta W|E\right).$$

Now we assume that offspring inherit their parents' fitnesses faithfully on average so the second, transmission effects, term is 0. We also note that since W is by definition the random variable for parental fitness in the parental environment, then $W|E = W$, so $\text{Cov}(W|E, W) = \text{Var}(W)$. Moving the constant (for each generation) $1/\text{E}(W)$ inside the variance enables us to arrive at equation (3.6) in the main text.

22. Steve Frank also relates the Price equation to geometrical views of selection acting to move populations, and in particular acting so as to accumulate *Fisher information* within populations, thereby further linking it with Fisher's earlier work [Frank, 2012].

23. The value of the fundamental theorem has been further considered by Samir Okasha in [Okasha, 2008] and by Jonathan Birch in [Birch, 2013], while A.W.F. Edwards considered its history, as well as its relationship to the work of Robertson and Price [Edwards, 2013].

24. Note that the heritability coefficient h^2 should be interpreted as heritability, rather than heritability squared; the h^2 notation is a historical artefact from Sewall Wright's presentation of it as a *path coefficient* squared [Wright, 1921b]. Heritability is *narrow-sense heritability* defined as

$$h^2 = \frac{\text{Var}(G)}{\text{Var}(P)}, \tag{3.29}$$

where $\text{Var}(G)$ is additive genetic variance of the trait in question and $\text{Var}(P)$ is its total phenotypic variance [Falconer and Mackay, 1996]. The definition of heritability in equation (3.29) makes clear that heritability measures the proportion of total phenotypic variation that is explainable by additive genetic variation. This agrees with the path coefficient interpretation above; since path coefficients are correlation coefficients, heritability is the square of a correlation coefficient. Most readers will recognize the correspondence with the R^2 statistic, or *coefficient of determination* [Grafen and Hails, 2002], which gives the proportion of variation explained by a linear regression.

25. Note that equation (3.8) is not presented as an estimator, as would be the case if this were a statistical regression given some data points. Rather, equation (3.8) gives us a way of rewriting the random variable G in terms of a linear function of the random variable P, and deviations from that linear function. The sum of the squares of these deviations, the residuals ε_{PG}, has been minimized by calculating the linear function using the least-squares approach (e.g., [Gardner et al., 2011]). In other words, equation (3.8) is simply a mathematical identity.

26. We know from the Price equation (3.3) that, assuming faithful transmission of traits on average, the total genetic change in a population is

$$\Delta \text{E}(G) = \text{Cov}\left(G, \frac{W}{\text{E}(W)}\right). \tag{3.30}$$

Substituting in the regression equation for G, (3.8), and expanding, since covariances are additive (i.e., $\text{Cov}(X + Y, Z) = \text{Cov}(X, Z) + \text{Cov}(Y, Z)$), we have

$$\Delta \text{E}(G) = \text{Cov}\left(\alpha_{GP}, \frac{W}{\text{E}(W)}\right) + \text{Cov}\left(P, \frac{W}{\text{E}(W)}\right)\beta_{GP} + \text{Cov}\left(\varepsilon_{GP}, \frac{W}{\text{E}(W)}\right)$$
$$= \text{Cov}\left(P, \frac{W}{\text{E}(W)}\right)\beta_{GP} + \text{Cov}\left(\varepsilon_{GP}, \frac{W}{\text{E}(W)}\right), \tag{3.31}$$

since the covariance of a constant (α_{GP}) with anything is 0. Then, if the residuals of the regression equation (3.8) are uncorrelated with fitness so that $\text{Cov}(\varepsilon_{GP}, W/\text{E}(W)) = 0$, the preceding equation simplifies to equation (3.9) in the main text.

27. Note that *narrow-sense heritability* is described in equation (3.9) as a regression of individual genotype on phenotype. This is the same as the resemblance between parents and offspring, illustrated in figure 3.2, since heritability defined according to equation (3.29) also corresponds to the regression of individual genetic value on expressed phenotype. This can be seen since phenotypic variance can be decomposed into additive genetic variance and the residual variance due to other factors, such as (genetic) environment, as [Falconer and Mackay, 1996]

$$\text{Var}(P) = \text{Var}(G) + \text{Var}(R). \tag{3.32}$$

Since G and R are necessarily uncorrelated, we can calculate

$$\text{Cov}(G, P) = \text{Cov}(G, G + R) = \text{Var}(G), \tag{3.33}$$

meaning that [Falconer and Mackay, 1996]

$$\beta_{GP} = \frac{\text{Cov}(G, P)}{\text{Var}(P)} = \frac{\text{Var}(G)}{\text{Var}(P)} = h^2. \tag{3.34}$$

28. Ensuring that the logical structure of the probabilistic model analyzed adequately captures causality is a problem of ensuring that a sufficient and appropriate set of predictors of fecundity are included in its construction [Queller, 2011, Frank, 2012]. One approach to analyzing such models in this way is *path analysis* as developed by Sewall Wright [Wright, 1921a, Shipley, 2002]. Similarly, *contextual analysis* uses group-level properties as predictors, in an attempt to determine the causal contribution of these to observed evolutionary change [Heisler and Damuth, 1987, Okasha, 2006].

29. Fisher's observations were motivated by the actuarial example of the contributions to population growth that individuals of different ages can be predicted to make; a young adult population member is much more likely to reproduce and contribute to population growth than a senescent individual, for example. This is discussed in more detail in chapter 10.

30. Grafen also provides a rigorous basis for the focus on arithmetic mean number of offspring in most applications of the Price equation, even in the face of arbitrary sources of uncertainty such as variability in offspring number, or environmental stochasticity, given a small number of simple and realistic assumptions [Grafen, 2000].

CHAPTER 4. INCLUSIVE FITNESS AND HAMILTON'S RULE

1. Hamilton's original papers argued that individual organisms should act as if to maximize their inclusive fitness [Hamilton, 1964a, Hamilton, 1964b]; we shall consider theoretical justifications for this claim in chapter 8.

2. Thus individuals interacting in pairs would experience the payoff matrix of the first social game we encountered in chapter 2, described by table 2.1.

3. The simple regression coefficients describing effects of own genetic values are the same as the more complicated partial regression coefficients we will see in the next chapter, when the random variables on which we are regressing are treated as independent. The linear partial regression coefficient arrived at using the least-squares method can be defined in terms of regression and correlation coefficients (e.g., [Gardner et al., 2011]) as

$$\beta_{XY|Z} = \frac{\beta_{XY} - \beta_{XZ}\beta_{ZY}}{1 - \rho_{YZ}^2}, \tag{4.8}$$

where $\beta_{XY|Z}$ denotes the partial regression of X on Y, holding Z constant, and ρ_{YZ} is the correlation coefficient of Y and Z, defined as

$$\rho_{YZ} = \frac{\text{Cov}(Y, Z)}{\sqrt{\text{Var}(Y)\text{Var}(Z)}}. \tag{4.9}$$

When Y and Z are independent then $\text{Cov}(Y, Z) = 0$ and thus $\beta_{ZY} = 0$ also, in which case equation (4.8) simplifies to

$$\beta_{XY|Z} = \beta_{XY}. \tag{4.10}$$

4. Genetic value is a linear function of underlying genes; this might be as simple as a 1 or 0 for possession or nonpossession of a gene, or the number of doses of an allele, such as 0, 1/2, or 1 for a single-locus diploid trait, or as complicated as breeding value estimated from random breeding into a reference population (see chapter 3).

5. Since we also need to account for baseline fitness due to nonsocial effects, a full description of fitness would decompose it into three components:

$$W = W_O + W_G + W_{G'}, \tag{4.11}$$

where W_O is baseline fitness independent of any social behavior by the individual or their social partners, W_G is the focal individual's effect on its own fitness, and $W_{G'}$ is the effect of the behavior of social partners on the focal individual's fitness [Queller, 1992b]. Queller proposed that effects on the focal individual's fitness be modeled as simple linear regressions. Since X given Y can be rewritten as

$$X = \alpha_{XY} + \beta_{XY}Y + \varepsilon_{XY}, \tag{4.12}$$

using such descriptions of fitness effects we can thus write out a full expression for fitness:

$$W = W_0 + \alpha_{WG \perp G'} + \beta_{WG \perp G'}G + \varepsilon_{WG \perp G'} + \alpha_{WG' \perp G} + \beta_{WG' \perp G}G' + \varepsilon_{WG' \perp G}. \tag{4.13}$$

There are three things worth noting about equation (4.13). First, although Queller proposed phenotype as the predictor in the regression models of fitness effects [Queller, 1992b], here for our simplest model it is sufficient to use genetic value (so there is no uncertainty or conditionality in the expression of social behavior, an assumption we will relax in subsequent chapters). Second, we notate the regression coefficients and other parameters of the fitness models using our notation for regression assuming statistical independence as in equation (4.2); this makes clear that these regressions capture only those fitness effects due to own or partners' genetic values (see note 8). Third, when we come to examine the residuals in the next chapter, these residuals will simply be the difference between the fitness due to own behavior (in the case of $\varepsilon_{WG \perp G'}$) and the linear regression model, or the difference between fitness due to social partners' behavior (in the case of $\varepsilon_{WG' \perp G}$) and the linear regression model. Failure to exclude other fitness effects would always make the residuals correlated, since G and G' will typically be correlated (cf. [Queller, 1992b]).

Now, in using equation (4.13) in the Price equation we can ignore the baseline fitness and the intercepts, and if the residuals are uncorrelated with W then these can be ignored as well (see note 6). Hence when residuals are uncorrelated, equation (4.13) simplifies to equation (4.2) in the main text.

6. When we use equation (4.13) in the Price equation, we wish to calculate $\text{Cov}(G, W)$, which gives us

$$\text{Cov}(G, W) = \text{Cov}(G, W_0)$$
$$+ \text{Cov}(G, \alpha_{WG \perp G'}) + \beta_{WG \perp G'}\text{Var}(G) + \text{Cov}(G, \varepsilon_{WG \perp G'}) \tag{4.14}$$
$$+ \text{Cov}(G, \alpha_{WG' \perp G}) + \beta_{WG' \perp G}\text{Cov}(G, G') + \text{Cov}(G, \varepsilon_{WG' \perp G}).$$

Baseline fitness W_0 is defined as fitness independent of any social behavior, hence is uncorrelated with the individual's genetic value making the first term 0. As α_{WG} and $\alpha_{WG'}$ are constants, their covariance with anything is 0, so if the residuals from the regression models of fitness based on own and partners' genetic values are uncorrelated with genetic value G, then the fourth and seventh terms are also 0. Since, when plugged into the Price equation, terms involving baseline fitness, the intercepts, and the residuals are all going to disappear (assuming residuals are uncorrelated with the focal individual's fitness), we can ignore them in equation (4.13) and simplify it to equation (4.2) in the main text (cf. chapter 3, note 26 and [Queller, 1992b]).

7. Neighbor-modulated fitness has also historically been referred to as direct fitness (e.g., [Taylor and Frank, 1996, Taylor et al., 2007b]), but given the potential for confusion with direct fitness as used in this book, which is a component of inclusive fitness, neighbor-modulated fitness is preferred.

8. To calculate the fitness regression coefficients we calculate β_{WG} and $\beta_{WG'}$ for equation (4.3), assuming G and G' to be independent (see note 3). From the definition of the regression coefficient (chapter 3, note 12) this gives us

$$\beta_{WG \perp G'} = \frac{\text{Var}(G)}{\text{Var}(G)}(-c) + \frac{\text{Cov}(G, G')}{\text{Var}(G)}b \tag{4.15}$$

and

$$\beta_{WG' \perp G} = \frac{\text{Cov}(G, G')}{\text{Var}(G')}(-c) + \frac{\text{Var}(G')}{\text{Var}(G')}b, \tag{4.16}$$

which, since if G and G' are independent then $\text{Cov}(G, G') = 0$, simplify to $\beta_{WG \perp G'} = -c$ and $\beta_{WG' \perp G} = b$, respectively.

9. Similarly to the additive donation game (note 8), for the additive public goods game (4.4) we calculate the simple regression coefficients β_{WG} and $\beta_{WG'}$ under the assumption that G and G' are independent. This gives

$$\beta_{WG \perp G'} = \frac{\text{Var}(G)}{\text{Var}(G)}\left(-c + \frac{b}{N}\right) + \frac{\text{Cov}(G, G')}{\text{Var}(G)}\frac{b}{N} \tag{4.17}$$

and

$$\beta_{WG' \perp G} = \frac{\text{Cov}(G, G')}{\text{Var}(G')}\left(-c + \frac{b}{N}\right) + \frac{\text{Var}(G')}{\text{Var}(G')}\frac{b}{N}, \tag{4.18}$$

which simplify to $\beta_{WG \perp G'} = -c + b/N$ and $\beta_{WG' \perp G} = b/N$, respectively.

10. Since covariances are additive; that is, even for correlated random variables, $\text{Cov}(X, Y + Z) = \text{Cov}(X, Y) + \text{Cov}(X, Z)$.

11. That is, $\text{Cov}(G, G') = \text{Cov}(G', G)$.

12. Our assumption that $\beta_{WG' \perp G} = \beta_{W'G \perp G'}$ means we assume that all individuals are of the same class, and do not differ in their ability or tendency to have a social effect on others' fitnesses. In chapter 7 we will see a version of Hamilton's rule that allows for such differences; the switch of perspective between neighbor-modulated and inclusive fitness forms of fitness will also be considered further in the next chapter. Jonathan Birch gives a detailed consideration of when the equivalence between inclusive and neighbor-modulated fitness formulations holds [Birch, 2013].

13. Under certain circumstances the correlation coefficient (Wright's *coefficient of relation-ship* [Wright, 1922, Hamilton, 1964a]) and regression coefficient (Hamilton's *coefficient of relatedness* [Hamilton, 1972]) formulations are equivalent. From the definition of correlation coefficient (see chapter 3, note 3) we can see that the two will be the same when $\text{Var}(G') = \text{Var}(G)$, since then

$$\frac{\text{Cov}(G, G')}{\sqrt{\text{Var}(G)^2}} = \frac{\text{Cov}(G, G')}{\text{Var}(G)}. \tag{4.19}$$

When variances are taken across the whole population, as they are under the Price-equation approach, and when focal individuals and social partners are drawn from the same population, then the $\text{Var}(G') = \text{Var}(G)$ condition may frequently be satisfied. In cases where the variances differ, however, the correlation measure of relationship will differ from the correct regression measure of relatedness (see, for example, [Falconer and Mackay, 1996, chapter 9]).

14. Identity-by-descent is defined as two alleles having been replicated from the same single ancestor allele in an earlier generation [Falconer and Mackay, 1996]. Identity-by-descent was used in early population genetical models that developed the basic logic of inclusive fitness theory (see note 15).

15. We consider the calculation of r from pedigree analysis as done using inbreeding coefficients [Falconer and Mackay, 1996]. Let $f_{GG'}$ represent the *coefficient of consanguinity* between individuals G and G', defined as the probability that two gametes, one taken at random from each individual, will contain alleles that are identical by descent (see note 14). Similarly, let f_G be the *coefficient of inbreeding*, defined as the probability that two alleles within the same individual are identical by descent. [Hamilton, 1972] derived the regression form of r as

$$r = \frac{2f_{GG'}}{1 + f_G}. \tag{4.20}$$

There are two things to note from this definition. First, the definition assumes fair meiosis and *autosomal* (i.e., non-*sex-linked* traits), since the factor 2 in the fraction arises from an assumption that a particular allele has a 50% chance of making it into a successful gamete. Second, to be able to calculate the pedigree relatednesses described by Hamilton, such as relatedness of 1/2 between siblings, the population needs to be *outbred* for the allele in question. This is because, in a population that is outbred for a particular allele, $f_{GG'} = 1/4$ for siblings, while $f_G = 0$ [Falconer and Mackay, 1996], which when substituted into equation (4.20) gives $r = 1/2$ for siblings as expected. The assumption of outbreeding for an allele is the assumption that the allele is rare, and consequently that selection on the allele is weak (otherwise the allele would rapidly increase in frequency). Simple pedigree analysis of r, of the kind originally described by [Hamilton, 1964a], thus requires weak selection.

16. John Seger subsequently related the different versions of the relatedness coefficient using a Price-equation approach, and also presented an apparently more general version that subsumed them [Seger, 1981].

17. Estimation of relatedness from genetic data relies on identifying marker alleles that are not subject to selection, then using these alleles in estimating the genetic regression coefficient. Such markers include alleles for *alloenzymes*, variants of enzymes with a common biological function but each having a differing amino acid sequence due to being coded for by a different allele [Queller and Goodnight, 1989].

18. For immediate responses to [Nowak et al., 2010] see [Abbot et al., 2011, Boomsma et al., 2011, Ferriere and Michod, 2011, Herre and Wcislo, 2011, Strassman et al., 2011]. For Nowak et al.'s reply see [Nowak et al., 2011]. Most criticisms of the original paper did not discuss the mathematical model of the evolution of eusociality advanced by Nowak, Tarnita, and Wilson; exceptions to this include a note that that model does not capture the key biological aspect of eusociality in many species, i.e., individual production of sexual males by unfertilized haplodiploid workers [Marshall, 2011a], and also a detailed analysis of the same model by Xiaoyun Liao, Stephen Rong, and David Queller, showing that it supports rather than refutes standard inclusive fitness predictions [Liao et al., 2015].

19. Philosophers of biology have investigated the potential differences and similarities between multilevel and inclusive fitness formulations. Jonathan Birch considers that the approaches differ in how they divide populations into groups, and has discussed difficulties with achieving the required grouping under either approach [Birch, 2013]. Using path analysis Samir Okasha has shown how, while inclusive fitness and multilevel selection models of the same scenario give equivalent predictions, they may differ in the causal explanations they suggest [Okasha, 2014].

20. To briefly explain the law of total covariance we adapt the proof of the *law of total variance* from [Weiss, 2006] as follows.

We start with the definition of the *law of total expectation* as (e.g., [Weiss, 2006])

$$E(X) = \sum_y E(X|Y = y)P(Y = y) = E(E(X|Y)), \tag{4.21}$$

which is an intuitive decomposition of expectation into its constituent expectations given different states of the world y. Applying this definition to the covariance between two random variables X and Y we can write

$$Cov(X, Y) = E[(X - E(X))(Y - E(Y))] = E[E((X - E(X))(Y - E(Y))|Z)]. \tag{4.22}$$

From the linearity of (conditional) expectation (i.e., $E(X + Y) = E(X) + E(Y)$) we can multiply out inside the expectation on the right-hand side of (4.22), and write this in full as

$$E[E(XY|Z) - E(X)E(Y|Z) - E(Y)E(X|Z) + E(X)E(Y)]. \tag{4.23}$$

Noting that $E(XY) = Cov(X, Y) + E(X)E(Y)$ we can in turn rewrite (4.23) as

$$E[Cov(X, Y|Z)] - E[E(X)E(Y|Z) - E(X|Z)E(Y|Z) + E(Y)E(X|Z) - E(X)E(Y)]. \tag{4.24}$$

The second term in this expression can then be factorized and recognized as a covariance between conditional expectations, giving us the law of total covariance:

$$Cov(X, Y) = E(Cov(X, Y|Z)) + Cov(E(X|Z), E(Y|Z)). \tag{4.25}$$

21. Price observed that the covariance equation describing selection at a given level could be inserted into the expectation term describing transmission bias at the immediately higher level of

selection [Price, 1972a]. Thus the approach is recursive in that transmission bias at one level of selection is described in terms of selection at the level immediately below. This recursion can be repeated indefinitely [Okasha, 2006].

22. Hamilton noted the apparent lack of an explicit definition of classical fitness ([Hamilton, 1996, p. 49], cited in [Segerstrale, 2013, p. 88]).

23. Peter Taylor and Steve Frank showed how evolutionary stability analysis techniques based on maximizing differential equations can be applied to social interactions [Taylor and Frank, 1996]. Their approach relies on mutations of small effect, and hence subject to weak selection, in order to be able to approximate partial derivatives of individual's and partners' phenotypes, with respect to individual's genetic value, using the corresponding regression coefficients. However, this assumption of weak selection is simply made in order to facilitate evolutionary stability analyses using derivatives; Taylor and Frank showed how, for mutations of larger effect, their approach relates to the partial regression formulation of inclusive fitness developed by David Queller and described in chapter 5. Note that in developing their approach, Taylor and Frank also argued that the neighbor-modulated view of inclusive fitness, initially proposed by Hamilton and used in the main text for this chapter, is typically easier to construct models with. For further discussion see chapter 10.

CHAPTER 5. NONADDITIVE INTERACTIONS AND HAMILTON'S RULE

1. To show that assortment rate α in the replicator dynamics model described by equations (5.1) and (5.2) is equivalent to the regression coefficient $\beta_{G'G}$ (see inequality (4.6)), and hence is a measure of relatedness, we simply calculate the regression coefficient for that model. Since individuals are either genetic donors (e.g., $G = 1$) or are not (e.g., $G = 0$), we can calculate $E(G^2) = E(G) = E(G') = f$ and

$$E(GG') = f(\alpha + (1 - \alpha)f). \tag{5.13}$$

Putting this all together we can then show that

$$\begin{aligned}
\beta_{G'G} &= \frac{\mathrm{Cov}(G, G')}{\mathrm{Var}(G)} \\[2mm]
&= \frac{E(GG') - E(G)E(G')}{E(G^2) - E(G)^2} \\[2mm]
&= \frac{\alpha f(1 - f)}{f(1 - f)} \\[2mm]
&= \alpha
\end{aligned} \tag{5.14}$$

(see also [Queller, 1984]).

2. We substitute equations (5.1) and (5.2) into the replicator dynamics (2.4). This gives us the replicator dynamics for the nonadditive donation game played between relatives as

$$\begin{aligned}
\frac{df}{dt} &= f\big(-c + \alpha(b + d) + (1 - \alpha)f(b + d) \\
&\quad - \big(f(-c + \alpha(b + d) + (1 - \alpha)f(b + d)) + (1 - f)(1 - \alpha)fb \big) \big).
\end{aligned} \tag{5.15}$$

3. The equilibrium is found by determining when equation (5.15) equals 0 (see chapter 2, note 10). This can be seen to be when

$$f = \frac{\alpha(b + d) - c}{(\alpha - 1)d} \tag{5.16}$$

(see also [Grafen, 1979] and [Queller, 1984] for equivalent results, although note that Queller introduces a sign error in presenting Grafen's result). We can then ask when this equilibrium can exist as a mixed-population equilibrium, that is, when the f specified by (5.16) is between 0 and 1. Solving for these constraints in terms of α, assuming a donation game ($b > 0$) where a negative deviation from additivity can never offset the benefit provided by donation (so $b + d > 0$), gives conditions (5.3) and (5.4) in the main text, for $d > 0$ and $d < 0$, respectively.

4. In deriving the conditions for the existence of the interior equilibrium, we assumed that $b > 0$ and $b + d > 0$ (see note 3). To check the stability of the equilibrium we differentiate the replicator dynamics with respect to f, then evaluate the sign of the derivative at the equilibrium point (see chapter 2, note 12). Differentiating equation (5.15) with respect to f then evaluating at (5.16) gives

$$\frac{(\alpha b - c + d)(\alpha(b + d) - c)}{(\alpha - 1)d}. \tag{5.17}$$

We next check the sign of expression (5.17) to determine the stability of the equilibrium; if (5.17) is positive then the equilibrium is unstable, whereas if it is negative then the equilibrium is stable (see chapter 2, note 12). Note that if the interior equilibrium exists then if $d > 0$ so α must be greater than $(c - d)/b$ but less than $c/(b + d)$ (by condition (5.3)), whereas by condition (5.4) the converse restrictions on α hold when $d < 0$; these facts, together with the assumptions that $b > 0$ and $b + d > 0$ (see note 3), mean that whenever the interior equilibrium exists, $b + d - c > 0$ must hold.

In expression (5.17) some algebra shows that, for a value of α that satisfies condition (5.3) or (5.4) (according to the sign of d), the numerator is always negative provided that $b + d - c > 0$, which we just established. The proof of this is left as an exercise to the interested reader. The sign of the denominator $(\alpha - 1)d$ in expression (5.17) clearly depends only on d, since necessarily $\alpha - 1 < 0$ as α cannot exceed 1; thus when $d < 0$ the denominator is positive, and vice versa. Given that, as described above, the numerator is always negative provided that the mixed equilibrium exists, the sign of the whole expression is thus determined by the sign of d, being negative when $d < 0$ and positive when $d > 0$. Thus, when nonadditivity is negative ($d < 0$) then the mixed-population equilibrium is stable, and when nonadditivity is positive ($d > 0$) it is unstable (see chapter 2, note 12).

5. This scenario is also considered by Steve Frank [Frank, 1998].

6. We consider the replicator dynamics acting independently at two loci, one for each behavioral role. We let the frequency of donation (type **I**) behavior for one role be f, and that for the other be g. Then the fitnesses of each strategy for each role are given by the equations

$$w_{\mathrm{I}f} = -c + g\left((1 + \alpha)(b + d)\right) + (1 - g)\alpha b, \tag{5.18}$$

$$w_{\mathrm{II}f} = gb, \tag{5.19}$$

$$w_{\mathrm{I}g} = -c + f\left((1 + \alpha)(b + d)\right) + (1 - f)\alpha b, \tag{5.20}$$

$$w_{\mathrm{II}g} = fb. \tag{5.21}$$

These can then be substituted into the replicator dynamics for f and g, respectively; analysis of these, assuming that $b + d > 0$ (cf. note 3) gives conditions (5.6) and (5.5) in the main text [Marshall, 2009].

7. Robert Axelrod and Bill Hamilton first proposed that inclusive fitness might explain the establishment, over evolutionary time, of reciprocal cooperation between unrelated individuals. Reciprocal cooperation will be discussed in a little more detail in the next chapter; for now it is sufficient to know that it provides a mechanism where, by conditioning social behavior on the previous outcomes of interactions with another individual, mutually beneficial social behavior may become established in a population even when interactions are not between relatives. Axelrod and Hamilton wrote,

> Once the genes for cooperation exist, selection will promote strategies that base cooperative behavior on cues in the environment.... When a cooperative choice has been made,

one cue to relatedness is simply the fact of reciprocation of cooperation. Thus modifiers for more selfish behavior after a negative response from the other are advantageous whenever the degree of relatedness is low or in doubt. As such, conditionality is acquired, and cooperation can spread into circumstances of less and less relatedness. Finally, when the probability of two individuals meeting each other again is sufficiently high, cooperation based on reciprocity can thrive and be evolutionarily stable in a population with no relatedness at all. [Axelrod and Hamilton, 1981]

This verbal proposal has been partially formalized by Richard McElreath and Rob Boyd who calculated payoffs for the strategies Tit-for-Tat (TFT: always begin by behaving socially towards a new individual, then in each subsequent interaction behave as the other did in the previous interaction), and always-defect (ALLD: never behave socially). The value of conditional social behavior depends on how long the interaction between any two individuals is likely to last, which can be modeled using a geometric distribution with a per-interaction probability that another interaction will occur, ω. Axelrod referred to ω as the "shadow of the future" [Axelrod, 1984], as changing it changes the discounting that should be applied to possible future benefits from establishing mutually beneficial prosocial behavior by both parties (receiving a fitness change on each interaction of $b - c$, where $b > c$ in the donation game of table 2.1), versus the immediate gains (not paying the cost c in the donation game of table 2.1) that may be had by behaving nonsocially. McElreath and Boyd calculate the payoffs of TFT and ALLD as [McElreath and Boyd, 2008]

$$w_{\text{TFT}} = \overbrace{(\alpha + (1 - \alpha)f)}^{\text{probability of interacting with TFT player}} \frac{b - c}{1 - \omega} - \underbrace{(1 - \alpha)(1 - f)}_{\text{probability of interacting with ALLD player}} c, \tag{5.22}$$

$$w_{\text{ALLD}} = (1 - \alpha)fb, \tag{5.23}$$

where f is the frequency of Tit-for-Tat players in the population, and $(b - c)/(1 - \omega)$ is the expected payoff to a Tit-for-Tat player interacting with another Tit-for-Tat player, given a geometrically distributed number of interactions with continuation probability ω.

We can take McElreath and Boyd's equations to calculate the critical value of relatedness α, above which Tit-for-Tat has higher fitness, as

$$\alpha > \alpha_{\text{crit}} = \frac{(1 - \omega)c + f\omega(c - b)}{b - c\omega + f\omega(c - b)}. \tag{5.24}$$

Axelrod and Hamilton's proposal hinges on the relatedness required for conditional cooperation (exemplified by Tit-for-Tat) progressively reducing. If we ask how α_{crit} changes as the frequency of Tit-for-Tat players in the population increases, we find that

$$\frac{\partial \alpha_{\text{crit}}}{\partial f} = -\frac{(b - c)^2 \omega}{(p\omega(c - b) + b - c\omega)^2}, \tag{5.25}$$

which is always negative since w is positive by definition (being a probability), and square numbers are also always positive. Thus, as the frequency of Tit-for-Tat players in the population increases, the level of relatedness required for Tit-for-Tat to experience positive selection decreases. This matches Axelrod and Hamilton's suggestion that conditional cooperation could spread into situations of lower and lower relatedness. This analysis is for the additive donation game; chapters 2 and 4, however, showed us that evolutionary outcomes can change when nonadditive fitness effects are considered.

Robert Axelrod considered a defining feature of a successful strategy for conditional cooperation to be *provocability* [Axelrod, 1984], meaning that exploitation of one individual by another, which occurs when one individual donates during a particular interaction whereas the other does not, should trigger an immediate noncooperative response. Such a situation (one individual having previously donated, the other not having done so) gives rise to one of Maynard Smith and Parker's *uncorrelated asymmetries*, however (see main text). If the two individuals, as a consequence of experiencing different behavioral roles, use different loci to determine their behavior, then different evolutionary outcomes can obtain. Let us consider a single locus that determines what an individual should do after it donated on the last interaction with a social partner, but that partner did not. Let us also assume that the interaction is a nonadditive donation game, as in table 2.2, and ask what the critical degree of relatedness required for donation to be favored in response to nondonation is. We obtain this by solving $w_{\mathrm{I}f} > w_{\mathrm{II}f}$ for α for the previous example (equations (5.18) and (5.19)), which gives

$$\alpha > \alpha_{\mathrm{crit}} = \frac{c - gd}{b + gd} \tag{5.26}$$

(assuming $b + d > 0$ as elsewhere). Now, as before, we ask how this critical relatedness threshold changes as the frequency of social behavior under the other behavioral role changes in the population, which is

$$\frac{\partial \alpha_{\mathrm{crit}}}{\partial g} = -\frac{d(b + c)}{(b + dg)^2}. \tag{5.27}$$

We can see that, since $b + c > 0$ by definition, the direction of change in the critical relatedness threshold depends only on the sign of d; in particular, if there is negative nonadditivity ($d < 0$) then the more frequent nonsocial behavior in the other behavioral role becomes (lower g), the *lower* the relatedness threshold becomes above which social behavior in response to nonsocial behavior is favored. Thus, when nonadditivity is negative, the more frequent nonsocial behavior in one role becomes, the easier it is for a prosocial response to experience positive selection; this outcome, illustrated in figure 5.3, is the exact opposite of Axelrod and Hamilton's proposal [Marshall and Rowe, 2003, Marshall, 2009].

8. Although the evolutionary stable points are the same whether one models selection acting on strategies for both behavioral roles simultaneously, or independently, the evolutionary dynamics differ, meaning that the equilibrium that selection results in from a given initial population composition can be different under the two approaches [Osinga and Marshall, 2015].

9. Queller was primarily interested in when the quantitative genetics separation captured in the breeder's equation (3.7), between selection and heritability, can be achieved for models of social behavior. He showed that satisfaction of the separation condition (5.7) determines when this can be achieved. Here, however, we are asking the related question of when Hamilton's rule can be derived with simple regression models of fitness effects. In chapter 4 we saw that the separation condition also answers this question (chapter 4, note 6). Note that the separation condition has been misinterpreted as predicting when fitness effects are completely independent of both population structure and population frequencies of behavioral types, and its failure to do so has been interpreted as proof that the multilevel-selection and Hamilton's-rule approaches to analyzing the direction of selection are not equivalent [van Veelen et al., 2012] (as was demonstrated in chapter 4); this viewpoint is incorrect [Birch and Marshall, 2014].

10. To show that nonadditivity gives rise to correlated residuals, and hence violation of the separation condition, we first calculate the regression coefficients describing the effect of own and social partners' genetic values on the fitness of a focal individual, for the nonadditive donation game described in table 2.2. We have three choices of how to deal with the additional parameter d; either we can model it as being a modification to the cost of a social behavior, conditional on the genetic value of social partners, or we can model it as being a conditional part of the benefit to others arising from a social behavior, or we can split the parameter d to make it partially a conditional modification to the cost and partially a conditional modification to the benefit. All of these will violate the separation condition; here we illustrate this for the case where d is considered as a conditional modification to the cost of a behavior.

So, we model fitness as

$$W = G(-c + G'd) + G'b. \tag{5.28}$$

Calculating the regression of W on G, assuming G is statistically independent of G', we have

$$\beta_{WG \perp G'} = \frac{\text{Cov}(G, W)}{\text{Var}(G)}$$

$$= \frac{E(GW) - E(G)E(W)}{E(G^2) - E(G)^2} \tag{5.29}$$

$$= \frac{g(-c + gd) - g^2(-c + gd)}{g(1 - g)}$$

$$= -c + gd.$$

Now we need to calculate the residuals $\varepsilon_{WG \perp G'}$, defined as the realized fitness due to own genetic value (excluding fitness effects due to social partners; see chapter 4, note 5) minus that predicted by the underlying regression model with the regression coefficient we just derived. So

$$\varepsilon_{WG \perp G'} = W_G - (\alpha_{WG \perp G'} + G\beta_{WG \perp G'}), \tag{5.30}$$

where W_G is that portion of a focal individual's fitness due to their own social behavior (see chapter 4, note 5). The residuals are thus calculated as in table 5.1; from this we can calculate

$$\text{Cov}(G, \varepsilon_{WG \perp G'}) = \alpha d(g - 1)^2 g, \tag{5.31}$$

which is always 0 (across all population frequencies g) only when interactions are additive ($d = 0$) and/or interactions are between nonrelatives ($\alpha = 0$). Thus nonadditive interactions are indeed a primary cause of the "failure" of inclusive fitness models [Queller, 1984, Queller, 1985, Queller, 1992b]. Extensions to inclusive fitness models that deal with such nonadditivity are introduced in the main text.

G G'	Predicted fitness	Actual fitness	Residual	Probability
0 —	$\alpha_{WG \perp G'}$	0	$-\alpha_{WG \perp G'}$	$(1 - g)$
1 0	$\alpha_{WG \perp G'} - c + gd$	$-c$	$-\alpha_{WG \perp G'} - gd$	$g(1 - \alpha)(1 - g)$
1 1	$\alpha_{WG \perp G'} - c + gd$	$-c + d$	$-\alpha_{WG \perp G'} + d - gd$	$g(\alpha + (1 - \alpha)g)$

Table 5.1: Residuals when additivity is a conditional part of the cost of a social behavior. Values of the four possible residuals are calculated along with the probability of each. These can be used to calculate the covariance between residuals and fitness.

11. Queller initially proposed, for a simple haploid trait where a genetic value of 0 indicates a nondonator, while a value of 1 indicates a donator, that the synergistic term for the nonadditive donation game be $GG'd$ [Queller, 1985] (Queller actually considered potentially conditional phenotypes P and P', but the form of the synergistic coefficient is the same). The formulation of the synergistic term as a random variable D, as in equation (5.8), gives a simple and general way of dealing with nonadditivity that can be applied to any kind of interaction [Marshall, 2014], and was suggested in passing by Queller himself in considering a general approach to modeling correlated characters [Queller, 1992a]. Unlike Queller's original regression on the joint predictor GG' (or PP' for phenotypes), the more

general random variable D does obscure the dependence of the synergistic effect on partners' genotypes [Marshall, 2014]. A similar extension to Hamilton's rule has been developed by Jeff Smith, J. David van Dyken, and Peter Zee, and applied to social evolution in microbial populations [Smith et al., 2010].

12. The synergistic coefficient for the nonadditive donation game (table 2.2), assuming a genetic value 0 for nondonators and 1 for donators, is very simply

$$\beta_{DG} = \frac{\text{Cov}(G, D)}{\text{Var}(G)} = \frac{\text{Cov}(G, GG')d}{\text{Var}(G)} \tag{5.32}$$

(cf. [Queller, 1985]).

13. As described in the main text, the random variable D captures the deviation from additivity of fitness interactions. For the nonadditive public goods game of expression (2.12) we write a focal individual's fitness as

$$W = -Gc + \frac{b(1 - \delta^{G+G'})}{N(1 - \delta)}. \tag{5.33}$$

We already know for the linear public goods game that additive effects of own and partners' social behavior on fitness are respectively $-c + b/N$ and b/N (see discussion of equation (4.4) in chapter 4). Then we simply need to solve

$$W = -Gc + \frac{b(1 - \delta^{G+G'})}{N(1 - \delta)} = G\left(-c + \frac{b}{N}\right) + G'\frac{b}{N} + D \tag{5.34}$$

for D which, very intuitively, gives us

$$D = \frac{b(1 - \delta^{G+G'})}{N(1 - \delta)} - \frac{(G + G')b}{N}. \tag{5.35}$$

From this we could calculate the synergistic coefficient β_{DG} given a suitable probabilistic model for the formation and composition of groups. Note that here, as in equation (4.4), group size is a random variable, and hence could change on a group-by-group basis [Marshall, 2014].

14. The approach ultimately taken by Queller echoes the commentary made by Alan Grafen when Queller's synergistic rule was first published, suggesting that Hamilton's rule works when costs and benefits are "averaged" [Grafen, 1985b].

15. Partial regression coefficients are introduced, and defined, in chapter 4, note 3.

16. The derivation of inequality (5.10) begins with a full model of fitness predicted from own and partners' genetic values:

$$W = \alpha_W + G\beta_{WG|G'} + G'\beta_{WG'|G} + \varepsilon_W. \tag{5.36}$$

Since the regression intercept α_W is a constant, its covariance with anything is 0, and since partial regression models necessarily result in residuals that are uncorrelated with the predictors used in constructing them, such as G and G' in this instance [Queller, 1992a], when we calculate the covariance of fitness with focal individual's genetic value G we get

$$\text{Cov}(G, W) = \text{Var}(G)\beta_{WG|G'} + \text{Cov}(G, G')\beta_{WG'|G}. \tag{5.37}$$

If we ask when this equation is positive (i.e., genetic value positively covaries with fitness, hence the behavior it represents experiences positive selection) and then divide the resulting inequality through by $\text{Var}(G)$, we get inequality (5.10) in the main text.

17. Note that here we have presented absolute fitness effects, rather than changes in relative fitness as in [Gardner et al., 2011].

18. Note however that Queller's synergistic extension of Hamilton's rule can be mathematically easier to work with than Queller's partial regression formulation [Marshall, 2014].

CHAPTER 6. CONDITIONAL BEHAVIORS AND INCLUSIVE FITNESS

1. Considering the rest of the genome an allele finds itself in as part of the environment can be traced back to [Fisher, 1930], as the discussion of the "fundamental theorem of natural selection" in chapter 3 shows.

2. The label "implicitly conditional" seems appropriate since it is frequently not recognized that such conditionality need even be considered in inclusive fitness theory. David Queller noted, for example, that Hamilton's original definition of inclusive fitness seemed too broad since it did not account for possible differences in behavioral options and fitness payoffs available to individuals, and hence had given rise to the paradoxes that Queller himself was seeking to address [Queller, 1996].

3. Trivers referred to the evolution of "reciprocal altruism," however given the strict meaning of altruism as behavior involving a lifetime direct fitness cost (see table 2.3), used throughout this book, reciprocal cooperation is here used instead (see also [West et al., 2007b]). Hamilton also made the same point to Trivers, subsequently writing, "I still believe the reciprocal altruism that Trivers explained to me and which he so named in his pioneer paper was misnamed" ([Hamilton, 1996, p. 263]; see also [Segerstrale, 2013, p. 136]).

4. Tit-for-Tat was submitted by the game theorist Anatol Rapoport to two reciprocal cooperation "tournaments" run by Axelrod, and won both [Axelrod, 1984].

5. The probability that Tit-for-Tat players interact with like players is thus a form of relatedness (see chapter 5, note 1).

6. The conditions presented in [Marshall, 2011c] are general versions of Axelrod and Hamilton's condition for a population of TFT players to be stable against invasion by ALLD ([Axelrod and Hamilton, 1981, first inequality of equations 1]),

$$\omega > \frac{c-d}{b},\tag{6.3}$$

and Axelrod's condition for the proportion of TFT players in the population required for TFT to invade ALLD ([Axelrod, 1984, pp. 212–213]),

$$f > \frac{c}{d + \frac{\omega}{1-\omega}(b-c+d)}.\tag{6.4}$$

Since the donation game shown in table 2.1 can be mapped onto the standard prisoner's dilemma payoffs as $b = T - P$, $-c = S - P$, and $d = R - S - T + P$, these conditions can easily be confirmed against the original results.

7. The simplest greenbeard model we might think of could consider pairwise interactions between individuals having two possible genetic values: $G = 1$ indicating that an individual is a greenbeard, and $G = 0$ indicating that they are not. Then an individual expresses prosocial behavior ($P = 1$) if and only if it is a greenbeard, and its social partner is as well; this can be captured by specifying the phenotype (i.e., expressed behavior) of both individuals as $P = GG'$ and $P' = GG'$, since these are both 1 only when G *and* G' are. To calculate whether condition (6.2) in the main text is satisfied, and thus determine whether greenbeards experience positive selection, we need to calculate $\mathrm{Cov}(G, P)$ and $\mathrm{Cov}(G, P')$. Since $P = P' = GG'$, we have $\mathrm{Cov}(G, P) = \mathrm{Cov}(G, P')$. This simplifies condition (6.2) to the condition

$$\beta_{WP\perp P'} + \beta_{WP'\perp P} > 0,\tag{6.5}$$

which, since $\beta_{WP\perp P'} = -c$ and $\beta_{WP'\perp P} = b$ as described in the main text, is simply the condition

$$b > c.\tag{6.6}$$

In other words, our analysis shows that greenbeards experience positive selection, even in unstructured populations, whenever the benefits arising from prosocial behavior exceed the costs of expressing it. In chapter 7 we consider the interpretation of this result in terms of inclusive fitness theory.

8. Linkage disequilibrium can be defined as the covariance between variables representing genetic values at two distinct loci (e.g., [Gardner et al., 2007b]), which may arise through physical linkage on a chromosome, so the two loci are unlikely to be split apart during genetic recombination, or through selection on traits with nonadditive fitness interactions so that, for example, having one trait or the other has a negligible effect on fitness but having both is very beneficial [Falconer and Mackay, 1996]. For two loci for conditional donation towards bearers of a phenotypic marker (call these "donors"), and the phenotypic marker itself (call these "bluebeards" to distinguish them from pleiotropic greenbeards), this can easily be derived as

$$\text{Cov}(G, Q) = q\,(\text{P(donor|bluebeard)} - g). \tag{6.7}$$

In the equation above, G represents the genetic value for the conditional donation trait ($G = 0$ represents nondonor while $G = 1$, occurring at population frequency g, represents conditional donation to bluebeards), while Q represents genetic value for the bluebeard phenotypic marker, which is unconditionally expressed ($Q = 0$ indicates no marker while $Q = 1$, appearing at population frequency q, indicates the individual has a blue beard). The conditional probability P(donor|bluebeard) is the probability that an individual has the allele for conditional donation ($G = 1$) given they have a bluebeard ($Q = 1$), and is equal to the expectation of G given Q since $\text{E}(G|Q) = 1 \times q\,(1 \times \text{P}(G = 1|Q = 1) + 0 \times (1 - \text{P}(G = 1|Q = 1))) + 0 \times (1 - q)(\ldots)$. Clearly, when having a bluebeard gives no information about whether that individual is a conditional donator, then $\text{P}(G = 1|Q = 1) = g$, which means that $\text{Cov}(G, Q) = 0$ indicating that the donation and bluebeard traits are in *linkage equilibrium*; otherwise they are in linkage disequilibrium.

9. It can be shown [Marshall, 2011c] that since for the bluebeard scenario of note 8 an individual's phenotype is $P = GQ'$ and their social partner's phenotype is $P' = G'Q$, then the covariance between fitness and genetic value for donation within an unstructured population is

$$\text{Cov}(G, W) = gq\,((1 - g)\beta_{WP \perp P'} + (\text{P(donor|bluebeard)} - g)\beta_{WP' \perp P}). \tag{6.8}$$

If we ask when this equation is positive then, for the standard donation game of table 2.1 (with $b > c > 0$), our condition for conditional donation to express positive selection is

$$\frac{\text{P(donor|bluebeard)} - g}{1 - g} > \frac{c}{b}. \tag{6.9}$$

We might conclude from this equation that a stable intermediate level of conditional donation will arise, as [Marshall, 2011c] erroneously suggests, but this is not quite the end of the story. To complete the analysis we need to consider what selection the phenotypic marker trait experiences. We thus need to calculate $\text{Cov}(Q, W)$, which we can do from condition (6.2) by simply swapping G for Q throughout. Noting that $\text{Cov}(Q, P) = \text{Cov}(Q, GQ')$ and $\text{Cov}(Q, P') = \text{Cov}(Q, G'Q)$, we can calculate these covariances, similarly to those in [Marshall, 2011c], as $\text{Cov}(Q, G'Q) = gq\,(1 - q)$ and $\text{Cov}(Q, GQ') = gq\,(\text{P(bluebeard|donor)} - q)$. Substituting these into condition (6.2), along with the values of the partial regression coefficients, and rearranging gives

$$\frac{1 - q}{\text{P(bluebeard|donor)} - q} > \frac{c}{b}. \tag{6.10}$$

This condition tells us when the phenotypic marker experiences positive selection which, since we know that $b > c$ (and hence $c/b < 1$) and also that the fraction on the left-hand side is always at least 1 (since we assumed a positive linkage disequilibrium between the two traits to begin with, so $q < \text{P(bluebeard|donor)} \leq 1$), is always the case.

Since the phenotypic marker always experiences positive selection, we need to consider what happens to P(donor|bluebeard) in condition (6.9). When the phenotypic marker is rare, and occurs only within individuals that are also conditional donators (otherwise selection would not have favored conditional donation towards bearers of the marker in the first place), P(donor|bluebeard) is close to 1 and so, from condition (6.9), conditional donation to bluebeards always experiences positive selection since $c/b < 1$. However, as we just saw, given that the phenotypic marker always experiences positive selection, it will eventually be driven to fixation in the population, at which point P(donator|bluebeard) = g. When this is the case, conditional donation towards bluebeards experiences negative selection, since $0 < c/b$. This result shows that greenbeard donation is inherently unstable. Jay Biernaskie, Stuart West, and Andy Gardner have considered selection on greenbeards in more detail, including the question of whether they are intragenomic outlaws [Biernaskie et al., 2011].

10. The analysis of the classic greenbeard scenario in terms of unconditionally expressed behaviors with nonadditive fitness effects is presented in chapter 7.

11. Greenbeard traits can also be *facultatively harming* or *obligately harming* [Gardner and West, 2010].

12. Gardner and West also note that synergistic terms are not needed when greenbeards are analyzed as conditional traits [Gardner and West, 2010]. To see in general how nonadditivity can be moved between genotype–phenotype relationships and phenotype–fitness relationships, consider fitness effects on an individual defined in terms of their own phenotype (ignoring regression intercepts similarly to chapter 4, note 5) as

$$W_G = \beta_{WP\perp P'} P + \varepsilon_{WP\perp P'}, \tag{6.11}$$

and phenotypes described in terms of underlying genotypes (again ignoring regression intercepts) as, for example,

$$P = G\beta_{PG\perp G'} + \varepsilon_{PG\perp G'}. \tag{6.12}$$

Putting both of these equations together we see that the fitness of an individual due to their own behavior can be written as

$$W_G = G\beta_{WP\perp P'}\beta_{PG\perp G'} + \varepsilon_{WP\perp P'} + \beta_{WP\perp P'}\varepsilon_{PG\perp G'}. \tag{6.13}$$

From this we can see that residuals from one regression can be moved into the other; by doing so we are able to redefine one relationship, either genotype–phenotype or phenotype–fitness, as completely linear with no residuals or with residuals uncorrelated with G (thereby satisfying the separation condition; see equation (5.7)), by moving the nonlinearities into the residuals of the regression describing the other relationship. While for a particular biological case this may or may not make sense, it illustrates the general interchangeability of conditionally expressed behaviors and nonadditive fitness effects in abstract models of social behavior.

CHAPTER 7. VARIANTS OF HAMILTON'S RULE AND EVOLUTIONARY EXPLANATIONS

1. As discussed in chapter 5 (see note 11), Queller actually simultaneously introduced conditional expression of phenotypes *and* the synergistic coefficient. These two generalizations have been separated out, with conditional expression of phenotypes treated independently in HR2 (see figure 7.1) since, as described in chapter 6, note 12, synergism and conditionality provide complementary analytic approaches and either is sufficient to describe conditional and/or nonadditive interactions. It is also worth noting that Queller did not originally use the simple (independent) regression notation for fitness effects arising from social behavior, not introducing it until later [Queller, 1992b].

2. Note that conditionally expressed behavior may reflect control on the part of the individual expressing the behavior, or external control such as *manipulation* via a social partner. Effects on partners' phenotypes are considered in more detail in chapter 8.

3. See note 1 for a discussion of the genesis of HR2.

4. To make condition (7.1) even more general, we could derive its equivalent version for continuous phenotypes by using integrals over phenotypes rather than summations.

5. The numerator and denominator in condition (7.1) are both derived as averages, with a leading $1/N$ coefficient where N is the number of social interactions within the population (see equation (3.17)). However, since these coefficients cancel when the ratio is taken they are not shown in condition (7.1).

6. Nee thus achieved in a general way what Jeff Fletcher and Martin Zwick achieved for a particular choice of payoff values in the iterated donation game [Fletcher and Zwick, 2006], as summarized in chapter 6.

7. To move from the genotype–phenotype version of Hamilton's rule (HR2 in figure 7.1) to the geometric view version (HR4) we simply use the covariance identity we originally saw in chapter 3: equation (3.17). First we transform HR2 from its neighbor-modulated form into its inclusive fitness form, since $\mathrm{Cov}(G, P')\beta_{WP'\perp P} = \mathrm{Cov}(G', P)\beta_{W''P\perp P'}$ (see chapter 4, note 12). This gives us

$$\beta_{WP\perp P'} + \frac{\mathrm{Cov}(G', P)}{\mathrm{Cov}(G, P)}\beta_{W''P\perp P'}. \tag{7.2}$$

Now we use the aforementioned covariance identity (3.17) to rewrite the covariance ratio, giving

$$\beta_{WP\perp P'} + \frac{\frac{1}{N}\sum_P(G' - E(G'))P}{\frac{1}{N}\sum_P(G - E(G))P}\beta_{W''P\perp P'}. \tag{7.3}$$

Finally we cancel the $1/N$ factors in the equation above, and make the assumption that expected genetic value for social partners is the same as it is for actors $(E(G') = E(G))$, to rewrite the above equation as HR4 in figure 7.1. We have thus derived the geometric view of relatedness (HR4) from first principles, as well as proving that it is equivalent to HR2.

8. See also the discussion of "kind selection" in chapter 8 for the possibility of direct fitness benefits returning to a focal member of a species via incidental increases in fitness of members of a partner species.

9. Since relatedness under the standard Price-equation approach, $\beta_{G'G}$, is calculated across the entire population (see chapter 3, note 5), in an unstructured population of greenbeards relatedness is 0 since $\mathrm{Cov}(G, G') = 0$.

10. Under the simple (independent) regression coefficient formulation of fitness effects (HR1 in figure 7.1), the cost of bearing a greenbeard trait, $\beta_{WG\perp G'} = -c$, is 0 since, as described in the main text, in the nonadditive donation game describing greenbeard interactions, $-c = 0$. This means that the cost, and benefit, of interacting with other greenbeards is captured entirely in the synergistic payoff parameter d, which must be positive since as previously shown, benefit dispensed by greenbeards must be greater than the cost they incur if they are to experience positive selection (see chapter 6, note 7).

11. The partial regression formulation of fitness effects (HR3 in figure 7.1) derives the cost of being a greenbeard, $\beta_{WG|G'}$, for the additive donation game (table 2.1) with conditionally expressed phenotypes. Fitness under this approach can be written as

$$W = GG'b - GG'c, \tag{7.4}$$

where individuals' genetic values are 1 if they are greenbeard bearers, and 0 otherwise. When we calculate the partial regression of fitness on the focal individual's genetic value (i.e., the cost in HR3), we note that if there is no assortment between individuals, so G and G' are independent, then the partial regression coefficient equals the simple regression coefficient (see chapter 4, note 3). Hence it is sufficient to calculate the partial regression coefficient as

$$\begin{aligned}
\beta_{WG|G'} &= \beta_{WG} \\
&= \frac{E(G^2 G') - E(G)E(GG')}{E(G^2) - E(G)^2}(b - c) \\
&= \frac{g^2(1 - g)}{g(1 - g)}(b - c) \\
&= g(b - c),
\end{aligned} \tag{7.5}$$

where g is the population frequency of the greenbeard trait. Since $b > c$ in the donation game, as long as there are greenbeards in the population, the net effect on direct fitness of bearing a greenbeard trait is actually positive (i.e., the "cost" is negative).

12. Here we consider what happens under the partial regression variant of Hamilton's rule (HR3) when the roles of potential donor and potential recipient are separated. We assume that each individual in a potential interaction occupies a distinct role, and which individual occupies which role is independent of the genetic value of either. We can then write individual fitness as

$$W = YGG'b - (1 - Y)GG'c, \tag{7.6}$$

where Y is a random variable denoting the role an individual finds itself in (1 if it has the opportunity to donate, 0 if its social partner has the opportunity to donate). Since Y is uncorrelated with G and G', then its expectation can be moved outside any covariances that it is involved in since, for example, $E(GY) = E(G)E(Y)$. Thus the calculation of the partial regression coefficient $\beta_{WG|G'}$ proceeds along similar lines to note 11 to give

$$\beta_{WG|G'} = g(E(Y)b - (1 - E(Y))c). \tag{7.7}$$

Under the reasonable assumption that an individual is equally likely to find themselves in either role, so $E(Y) = 1/2$, this simplifies to

$$\beta_{WG|G'} = g\left(\frac{1}{2}(b - c)\right). \tag{7.8}$$

This equation shows that even if greenbeard donation, when it occurs, results in an uncompensated lifetime reduction in direct fitness, HR3 still classifies the trait as conferring a direct fitness advantage on average, since $b - c > 0$ and hence $(b - c)/2 > 0$. A very similar calculation of the synergistic coefficient β_{DG} from HR1 could be done for this scenario, with the same conclusion.

13. It is straightforward to derive the cost in HR2 and HR4 by starting from a description of fitness in terms of expressed phenotypes,

$$W = P'b - Pc, \tag{7.9}$$

where P and P' are the expressed phenotypes (1 for donation, 0 for nondonation) of the focal individual and their social partner(s), respectively. Then $\beta_{WP \perp P'} = -c$ is derived from this equation in exactly the same way as $\beta_{WG \perp G'} = -c$ is derived from an unconditional nonadditive donation game (see chapter 4, note 8). Since $c > 0$ in the donation game (table 2.1), the effect of donation on an individual's direct fitness is negative, and thus appears altruistic as discussed in the main text.

14. "Relatedness" in HR2 is defined as

$$\frac{\text{Cov}(G, P')}{\text{Cov}(G, P)}. \tag{7.10}$$

In this expression, G and G' are 1 if, respectively, the focal individual and their social partner(s) are greenbeards, and 0 otherwise. Then the expressed phenotypes (1 if an individual donates and 0 otherwise) can be written as $P = GG'$ and $P' = GG'$ since greenbeards donate only when interacting with other greenbeards. Substituting these expressions into the covariance ratio above simplifies it to 1. Since the genetic geometric relatedness in HR4 is simply a mathematical rewriting of this relatedness (see note 7) it too will be 1.

15. Jay Biernaskie and colleagues also identify the different approaches to analyzing greenbeards-as-altruists, and make similar remarks on the validity of the different resulting classifications [Biernaskie et al., 2011].

CHAPTER 8. HERITABILITY, MAXIMIZATION, AND
EVOLUTIONARY EXPLANATIONS

1. For a particularly lucid explanation of Hamilton's hopes for inclusive fitness theory, and how he conceived its relationship to the earlier work arising during the modern synthesis, especially that of Fisher, see [Segerstrale, 2013].

2. Arguments for differential survival or reproduction of groups as an evolutionary explanation for social behavior have a long history, but have most recently been exemplified in the *Nature* article by Martin Nowak, Corina Tarnita, and E. O. Wilson [Nowak et al., 2010], in which they explicitly claim that genetic relatedness (and hence inclusive fitness) does not have a causal role, but rather arises as a by-product of selection acting on groups. The claims by Nowak, Tarnita, and Wilson echo earlier calls by E. O. Wilson and Bert Hölldobler for group selection to replace inclusive fitness theory as an evolutionary explanation of altruism [Wilson and Hölldobler, 2005]. Wilson and Hölldobler's criticism of inclusive fitness theory led to several responses, for example by Kevin Foster, Tom Wenseleers, and Francis Ratnieks [Foster et al., 2006a]. Nowak, Tarnita, and Wilson also emphasize the importance of *preadaptations* to sociality; interestingly Hamilton made similar remarks several decades earlier ([Hamilton, 1972], discussed in [Segerstrale, 2013]). In the same article, as discussed in chapter 4, Nowak, Tarnita, and Wilson also claim that neighbor-modulated fitness can be the target of selection, with inclusive fitness relegated to a conceptually confusing alternative viewpoint.

3. Samir Okasha [Okasha, 2006] presents an analysis of Richard Lewontin's conditions for evolutionary change to take place (referred to as the *Lewontin conditions* [Lewontin, 1970]), in terms of the Price equation. The Lewontin conditions are threefold: [Okasha, 2006]

(i) Phenotypic variability ($\mathrm{Var}(P) \neq 0$)
(ii) Phenotype-associated differences in fitness ($\beta_{WP} \neq 0$)
(iii) Heritability ($\beta_{GP} \neq 0$)

It is worth remarking that Lewontin's original condition (iii) was for heritability of individual fitness, rather than phenotype; condition (iii) is thus a modification of Lewontin's original, due to Okasha [Okasha, 2006].

4. In fact, while the Lewontin conditions (see note 3) generally predict when evolutionary change will occur, they are not logically sufficient, since they can be satisfied in certain situations in which no evolutionary change occurs [Okasha, 2006]. For our purpose, however, which is to highlight the necessity of heritability for evolutionary change when there is no transmission bias, they are useful.

5. David Queller's approach to deriving a quantitative genetics viewpoint on multilevel selection starts with the decomposition of individual values into components corresponding to group mean and deviation from group mean [Queller, 1992b], such as

$$W = \overline{W} + W_\Delta \tag{8.11}$$

and

$$G = \overline{G} + G_\Delta. \tag{8.12}$$

Taking equations (8.11) and (8.12) and, assuming zero transmission bias, substituting them into the Price equation (3.3) gives

$$
\begin{aligned}
\mathrm{E}(W)\Delta\mathrm{E}(G) &= \mathrm{Cov}(G, W) \\
&= \mathrm{Cov}(\overline{G} + G_\Delta, \overline{W} + W_\Delta) \\
&= \mathrm{Cov}(\overline{G}, \overline{W}) + \mathrm{Cov}(\overline{G}, W_\Delta) + \mathrm{Cov}(G_\Delta, \overline{W}) + \mathrm{Cov}(G_\Delta, W_\Delta) \\
&= \mathrm{Cov}(\overline{G}, \overline{W}) + \mathrm{Cov}(G_\Delta, W_\Delta).
\end{aligned}
\tag{8.13}
$$

The last step of the simplification follows because deviations from group mean genetic value are necessarily uncorrelated with group mean fecundity ($\mathrm{Cov}(G_\Delta, \overline{W}) = 0$), and deviations from group mean fecundity are similarly uncorrelated with group mean genetic value ($\text{Cov}(W_\Delta, \overline{G}) = 0$). The resulting equation does not look quite like the Price equation version of multilevel selection derived in chapter 4 (equation (4.7)), but we can see that it is in fact identical, as follows. Taking the first term first, by applying the law of total expectation (see chapter 4, note 20) we can rewrite it as

$$\text{Cov}(\overline{G}, \overline{W}) = \text{E}(\overline{GW}) - \text{E}(\overline{G})\text{E}(\overline{W})$$

$$= \text{E}(\text{E}(\overline{GW}|I)) - \text{E}(\text{E}(\overline{G}|I)\text{E}(\text{E}(\overline{W}|I))) \tag{8.14}$$

$$= \text{Cov}(\text{E}(\overline{G}|I), \text{E}(\overline{W}|I)),$$

where I is a random variable that uniquely identifies each group. Since the expected values of the group means for a given group are simply the expectations of the individuals' values within a given group, we can further rewrite the resulting equation as

$$\text{Cov}(\text{E}(\overline{G}|I), \text{E}(\overline{W}|I)) = \text{Cov}(\text{E}(G|I), \text{E}(W|I)), \tag{8.15}$$

which shows the first term of Queller's covariance equation (8.13) above is equal to the between-group selection term we first derived in chapter 4. Turning our attention to the second term in equation (8.13) we can use the law of total covariance (see chapter 4, note 20) to rewrite it as

$$\text{Cov}(G_\Delta, W_\Delta) = \text{Cov}(\text{E}(G_\Delta|I), \text{E}(W_\Delta|I)) + \text{E}(\text{Cov}(G_\Delta, W_\Delta|I)), \tag{8.16}$$

where, as before, I is the random variable uniquely indexing each group. Since, for a particular group, the average of the within-group deviations from the group mean is necessarily 0, the first term in the right-hand side of this equation is 0. Then, since from equations (8.11) and (8.12) we have $G_\Delta = G - \overline{G}$ and $W_\Delta = W - \overline{W}$, we can rewrite the second term as

$$\text{E}(\text{Cov}(G_\Delta, W_\Delta|I)) = \text{E}(\text{Cov}(G - \overline{G}, W - \overline{W}|I))$$

$$= \text{E}(\text{Cov}(G, W|I)), \tag{8.17}$$

where the final simplification follows because, within a given group, $\overline{G}$ and $\overline{W}$ are constants, so do not contribute to the covariance.

Putting all of this together, we have thus shown that

$$\text{Cov}(\overline{G}, \overline{W}) + \text{Cov}(G_\Delta, W_\Delta) = \text{Cov}(\text{E}(G|I), \text{E}(W|I)) + \text{E}(\text{Cov}(G, W|I)). \tag{8.18}$$

In other words, Queller's formulation of multilevel selection (the left-hand side of the equation above) equals the version we derived in chapter 4 (the right-hand side of the equation). Queller's version is simpler to work with for our present purpose, however; to start with, we take its phenotypic equivalent,

$$\text{Cov}(\overline{P}, \overline{W}) + \text{Cov}(P_\Delta, W_\Delta), \tag{8.19}$$

and, under the quantitative genetics approach of assuming only phenotypes are directly observable, try to come up with an expression for $\text{Cov}(G, W)$ by modeling mean genetic values, and individuals' deviations from them, using simple linear regressions

$$\overline{G} = \alpha_{\overline{GP}} + \beta_{\overline{GP}}\overline{P} + \varepsilon_{\overline{GP}} \tag{8.20}$$

and

$$G_\Delta = \alpha_{G_\Delta P_\Delta} + \beta_{G_\Delta P_\Delta} P_\Delta + \varepsilon_{G_\Delta P_\Delta}. \tag{8.21}$$

Assuming the residuals from these regressions are uncorrelated with G (the separation condition, see equation (5.7)), these can be substituted into expression (8.19) to give equation (8.2) in the main text, as required.

6. Heritability of family means can be derived, in the notation used in this chapter, as [Falconer and Mackay, 1996]

$$\beta_{\overline{G}\overline{P}} = \frac{\mathrm{Var}(\overline{G})/\mathrm{Var}(G)}{\mathrm{Var}(\overline{P})/\mathrm{Var}(P)} \beta_{GP}$$

$$= \frac{\beta_{\overline{G}G}}{\beta_{\overline{P}P}} \beta_{GP},$$

(8.22)

where β_{GP} is individual-level heritability (see chapter 3, note 27) and $\beta_{\overline{G}G}$, for example is the regression of mean group genetic value on individual genetic value. Rewriting $\mathrm{Var}(\overline{G})/\mathrm{Var}(G)$ as $\beta_{\overline{G}G}$ can be done since, within a group, individual deviation from group mean genetic value is uncorrelated with group mean genetic value itself (as the interested reader could easily show for groups of size two, for example), so $\mathrm{Var}(\overline{G}) = \mathrm{Cov}(\overline{G}, \overline{G} + G_\Delta) = \mathrm{Cov}(\overline{G}, G)$, and similarly for the phenotypic regression coefficient. Note that this genetic regression coefficient is not quite the relatedness regression coefficient of Hamilton's rule, since the value of the focal individual is included in the calculation of the group mean; this means that even in completely randomly formed groups, heritability of group means is nonzero, due to the resemblance between the focal individual and its offspring, as one would expect.

We have thus shown that heritability of group means is simply a function of the within-group genetic resemblance, the within-group phenotypic resemblance, and individual-level heritability. This is particularly important since, as described in the main text, in the most stringent form of group selection in which individuals are selected to reproduce solely on the mean phenotypic value of their group, heritability still involves within-group measures of similarity and the individual-level measure of heritability.

For completeness, the heritability of within-group deviations can be derived as [Falconer and Mackay, 1996]

$$\beta_{G_\Delta P_\Delta} = \frac{1 - \beta_{\overline{G}G}}{1 - \beta_{\overline{P}P}} \beta_{GP}.$$

(8.23)

As we would expect, this heritability also depends on individual-level heritability.

7. Note that in discussing multilevel selection (MLS) we are implicitly referring to what Samir Okasha refers to as *MLS1*, in which groups do not reproduce but rather individuals within them do; in *MLS2*, groups do reproduce and so group-level heritability becomes meaningful [Okasha, 2006].

8. Our starting point for deriving a quantitative genetics separation of neighbor-modulated fitness is actually the equation for total evolutionary change,

$$\Delta E(G) \propto \mathrm{Cov}(G, P)\beta_{WP\perp P'} + \mathrm{Cov}(G, \varepsilon_{WP\perp P'}) + \mathrm{Cov}(G, P')\beta_{WP'\perp P} + \mathrm{Cov}(G, \varepsilon_{WP'\perp P}).$$

(8.24)

In the following we assume that the residuals from the independent regressions are uncorrelated with the focal individual's genetic value (i.e., the separation condition is satisfied; see equation (5.7)), in which case the second and fourth terms in the equation above are 0. We can then divide this simpler equation through by $\mathrm{Cov}(G, P)$ and ask when the resulting right-hand side is positive, in order to derive inequality (6.2) (HR2 in chapter 7). Working directly with the present equation, however, we observe that we can rewrite everything and then move variances (see chapter 3, note 26) to get [Queller, 1992b]

$$\Delta E(G) \propto \mathrm{Cov}(G, P)\frac{\mathrm{Cov}(W, P \perp P')}{\mathrm{Var}(P)} + \mathrm{Cov}(G, P')\frac{\mathrm{Cov}(W, P' \perp P)}{\mathrm{Var}(P')}$$

$$\propto \frac{\mathrm{Cov}(G, P)}{\mathrm{Var}(P)}\mathrm{Cov}(W, P \perp P') + \frac{\mathrm{Cov}(G, P')}{\mathrm{Var}(P')}\mathrm{Cov}(W, P' \perp P),$$

(8.25)

which is equivalent to equation (8.3) in the main text. Note that covariances involving fitness retain the assumption that the phenotypes of the focal individual and their social partners are independent (see chapter 4, note 3).

9. As shown in chapter 3, note 27; note that phenotypic resemblance between a social partner's offspring and the focal individual is *half* the heritability that would be derived if the focal individual were actually the parent, by regressing offspring phenotypic value on midparent phenotypic value [Falconer and Mackay, 1996].

10. To further examine the nature of neighbor-modulated and inclusive fitness heritability, and their relative valuation, we start with the neighbor-modulated quantitative genetics equation of (8.3) and divide through by β_{GP}, giving

$$\frac{\Delta E(G)}{\beta_{GP}} \propto \mathrm{Cov}(W, P \perp P') + \mathrm{Cov}(W, P' \perp P)\frac{\beta_{GP'}}{\beta_{GP}}$$

$$\propto \mathrm{Cov}(W, P \perp P') + \mathrm{Cov}(W, P' \perp P)\frac{\mathrm{Cov}(G, P')/\mathrm{Var}(P')}{\mathrm{Cov}(G, P)/\mathrm{Var}(P)} \qquad (8.26)$$

$$\propto \mathrm{Cov}(W, P \perp P') + \mathrm{Cov}(W, P' \perp P)\frac{\mathrm{Cov}(G, P')}{\mathrm{Cov}(G, P)},$$

where the final step follows from an additional assumption that phenotypic variance in potential donors and potential recipients is equal ($\mathrm{Var}(P) = \mathrm{Var}(P')$). Since β_{GP} is assumed to be positive this equation will be positive whenever total evolutionary change $\Delta E(G)$ is positive, so our condition for a social trait to experience positive selection is

$$\mathrm{Cov}(W, P \perp P') + \mathrm{Cov}(W, P' \perp P)\frac{\mathrm{Cov}(G, P')}{\mathrm{Cov}(G, P)} > 0. \qquad (8.27)$$

Proceeding similarly for the inclusive fitness formulation of equation (8.4),

$$\frac{\Delta E(G)}{\beta_{GP}} \propto \mathrm{Cov}(W, P \perp P') + \mathrm{Cov}(W', P \perp P')\frac{\beta_{G'P}}{\beta_{GP}}$$

$$\propto \mathrm{Cov}(W, P \perp P') + \mathrm{Cov}(W', P \perp P')\frac{\mathrm{Cov}(P, G')/\mathrm{Var}(P)}{\mathrm{Cov}(P, G)/\mathrm{Var}(P)} \qquad (8.28)$$

$$\propto \mathrm{Cov}(W, P \perp P') + \mathrm{Cov}(W', P \perp P')\frac{\mathrm{Cov}(G', P)}{\mathrm{Cov}(G, P)},$$

so the condition for a social trait to experience positive selection is

$$\mathrm{Cov}(W, P \perp P') + \mathrm{Cov}(W', P \perp P')\frac{\mathrm{Cov}(G', P)}{\mathrm{Cov}(G, P)} > 0. \qquad (8.29)$$

It is this version of HR2 that can be rewritten directly as the genetic, geometric view of relatedness (see section 7.2).

We have thus come full circle for both the neighbor-modulated and inclusive fitness versions of HR2. In earlier chapters we derived variants of Hamilton's rule, in neighbor-modulated and inclusive fitness forms, in terms of assortment between social partners. Here we have derived the same conditions, again in neighbor-modulated but also more importantly in inclusive fitness forms, through analyzing the relative heritabilities of social traits via direct offspring and via the offspring of social partners. As discussed in the main chapter text, this suggests that a distinction between assortment of social traits, and heritability of social traits, is artificial. Furthermore, as also discussed in the chapter text, under the inclusive fitness viewpoint the ratio of heritability via social partners' offspring to heritability via direct offspring, $\beta_{G'P}/\beta_{GP}$, gives the relative valuations of these types of offspring required by inclusive fitness theory, and since $\beta_{G'P}/\beta_{GP} = \mathrm{Cov}(G', P)/\mathrm{Cov}(G, P)$, this valuation is based on genetic similarity, according to the geometric view of relatedness, as shown in section 7.2.

11. Andy Gardner and John Welch have developed the maximization approach under arbitrary interactions, including nonadditive ones, but considering the gene rather than the individual as the maximizing agent [Gardner and Welch, 2011].

12. Lehmann and Rousset's paper is a contribution to a special issue of the journal *Biology and Philosophy* on the formal Darwinism project, including a target review by Alan Grafen [Grafen, 2014a], a number of commentaries including that of Lehmann and Rousset's, and a response by Grafen to these [Grafen, 2014b].

13. Tim Clutton-Brock has reviewed the evidence in favor of reciprocal cooperation existing in nonhuman animals, and concluded that apparent examples of reciprocation can be explained by alternative, simpler, mechanisms [Clutton-Brock, 2009]. For an alternative interpretation, however, see the subsequent review by Michael Taborsky [Taborsky, 2013].

14. Lehmann and Keller's framework appeared as a target review in the *Journal of Evolutionary Biology* [Lehmann and Keller, 2006b], and attracted broadly favorable comments from a number of leading inclusive fitness experts. Lehmann and Keller's response to this commentary is also its most effective summary [Lehmann and Keller, 2006a].

15. The derivation presented in [Marshall, 2011c] of Hamilton's rule for the graph-theoretic results of [Ohtsuki et al., 2006] makes use of approximations Ohtsuki and colleagues used to derive their own condition for the evolution of altruism on graphs, then substitutes them into the Price equation to show that Hamilton's rule with genetic relatedness exactly matches the Ohtsuki et al. condition. Laurent Lehmann, Laurent Keller, and David Sumpter took a population-genetics approach to the same end [Lehmann et al., 2007], as did Alan Grafen [Grafen, 2007b].

16. Nowak, Tarnita, and Wilson also, in the supplementary material for their 2010 *Nature* paper, recognize the potential for transmission bias to affect evolutionary outcomes [Nowak et al., 2010].

17. Here we use Queller's "kin selection" label, although it has been avoided throughout the book. Queller uses the label to cover interactions between individuals related by pedigree, and differentiates this from relatedness arising through other routes, such as occurs in greenbeard interactions. See chapter 4 for the historical origins of the phrase.

18. See chapter 4, note 12 on the switch between neighbor-modulated and inclusive fitness forms.

19. Jonathan Birch and James Marshall discuss a model by Matthijs van Veelen and colleagues [van Veelen et al., 2012], showing how probabilistic expression of phenotypes can interact with nonadditive fitness effects between relatives to violate the separation condition required for inequality (8.6) [Birch and Marshall, 2014].

20. As noted in chapter 5 (note 11) this permits the same rule to easily describe nonadditive interactions in groups larger than two.

21. Queller's approach is thus in the same tradition as *indirect genetic effects* models (e.g., [McGlothlin et al., 2014]), which are the social trait version of classic approaches to dealing with selection acting on correlated characters [Lande and Arnold, 1983].

22. Inequality (8.8) in the main text is derived by starting with the phenotypic equivalent of the standard partial-regression neighbor-modulated description of fitness, used to derive HR3, as

$$W = \beta_{WP|P'} P + \beta_{WP'|P} P' + \varepsilon_W. \tag{8.30}$$

Note that the residuals from this regression may correlate with focal individual's genetic value, G, since this is not among the predictors used in the partial regression (see section 5.2.2). When $\mathrm{Cov}(G, \varepsilon_W) = 0$ we can substitute a prediction of social partners' phenotype in terms of focal individual's phenotype (8.33) into the above equation and ask when $\mathrm{Cov}(G, W) > 0$, to give

$$\beta_{WP|P'} \mathrm{Cov}(G, P) + \beta_{WP'|P} \left(\mathrm{Cov}(G, P)\beta_{P'P} + \mathrm{Cov}(G, \varepsilon_{P'P}) \right) > 0. \tag{8.31}$$

When $\mathrm{Cov}(G, \varepsilon_{P'P}) = 0$ the above condition can be divided through by $\mathrm{Cov}(G, P)$ to give condition (8.8) in the main chapter text.

23. When the effect on partners' fitness arising from manipulation of their behavior is incidental, the behavior experiences positive selection because changing the partners' behavior feeds back to increase

the focal individual's fitness. The selection condition can be derived from condition (8.8) by noting that $\beta_{W''P} = \beta_{W''P'}\beta_{P'P}$ [Queller, 2011]. This can be confirmed by expressing the partners' fitness in terms of their own behavior (there being no direct influence of focal individual's behavior on partners' fitness, only an indirect influence via their effect on the partners' behavior) as

$$W'' = \alpha + \beta_{W''P'}P' + \varepsilon_{W''P'},\tag{8.32}$$

and expressing the partners' behavior in terms of the focal individual's behavior as

$$P' = \alpha + \beta_{P'P}P + \varepsilon_{P'P}.\tag{8.33}$$

Then we wish to calculate

$$
\begin{aligned}
\beta_{W''P} &= \frac{\operatorname{Cov}(W'', P)}{\operatorname{Var}(P)} \\
&= \frac{\beta_{W''P'}\operatorname{Cov}(P, P') + \operatorname{Cov}(P, \varepsilon_{W''P'})}{\operatorname{Var}(P)} \\
&= \frac{\beta_{W''P'}\left(\beta_{P'P}\operatorname{Var}(P) + \operatorname{Cov}(P, \varepsilon_{P'P})\right) + \operatorname{Cov}(P, \varepsilon_{W''P'})}{\operatorname{Var}(P)} \\
&= \beta_{W''P'}\beta_{P'P}.
\end{aligned}
\tag{8.34}
$$

The final step, giving the result we were seeking, follows because the covariances involving residuals are necessarily 0 in this case: $\operatorname{Cov}(P, \varepsilon_{P'P}) = 0$ because P is the predictor in the corresponding linear regression, hence the residuals cannot be correlated with it, while $\operatorname{Cov}(P, \varepsilon_{W''P'}) = 0$ because P has no direct effect on W'', only an indirect effect via P' which is, once again, the predictor in the corresponding linear regression and therefore uncorrelated with the residuals.

We rearrange the result as $\beta_{P'P} = \beta_{W''P}/\beta_{W''P'}$ and substitute this into condition (8.8) in the main text to give, after a simple rearrangement,

$$\beta_{WP|P'} + \beta_{W''P}\frac{\beta_{WP'|P}}{\beta_{W''P'}}.\tag{8.35}$$

Condition (8.35) shows that an individual only values the change in partners' fitness arising from its own behavior ($\beta_{W''P}$), to the extent that the modification to partners' behavior feeds back to direct fitness benefits for itself ($\beta_{WP'|P}$) [Queller, 2011].

24. David Queller's derivation of his second condition for kith selection [Queller, 2011] differs from that presented here, since it contains a number of errors. Queller's initial description of individual fitness contains a clear typographical error but, more importantly, Queller presents a condition for selection on individual behavior P rather than underlying genetic value G; Joel McGlothlin and colleagues present a similar phenotypic selection rule, but note that it does not explain genetic change [McGlothlin et al., 2014]. Here a simpler version in terms of regressions directly on genetic value is presented, as well as a corrected version of Queller's inequality for conditional behaviors.

Starting with the description of fitness given in the main text (8.9), we ask when genetic value positively covaries with reproduction, and calculate

$$
\begin{aligned}
\operatorname{Cov}(G, W) &= \beta_{WG|W''}\operatorname{Var}(G) + \beta_{WW''|G}\operatorname{Cov}(G, W'') + \operatorname{Cov}(G, \varepsilon_W) \\
&= \beta_{WG|W''}\operatorname{Var}(G) + \beta_{WW''|G}\operatorname{Cov}(G, W''),
\end{aligned}
\tag{8.36}
$$

with the last step coming since the residuals from a partial regression are necessarily uncorrelated with any of the predictors of that partial regression (see section 5.2.2). We then divide this expression through by $\operatorname{Var}(G)$ and ask when the resulting expression is positive, to give condition (8.10) in the main text.

To come up with the corresponding condition for conditional behaviors, we would instead start with the description of fitness

$$W = \beta_{WP|W'} P + \beta_{WW'|P} W' + \varepsilon_W, \tag{8.37}$$

then repeat the previously performed steps to arrive at the condition

$$\beta_{WP|W'} + \beta_{WW'|P} \frac{\text{Cov}(G, W')}{\text{Cov}(G, P)} > 0. \tag{8.38}$$

CHAPTER 9. WHAT IS FITNESS?

1. Although Haldane's drowning-relative question has been taken for evidence that the idea of the importance of genetic relatedness in explaining altruism predated Hamilton's insights, Haldane appears to have been using this scenario to consider the necessity of small population sizes for rare genes to spread under positive selection; the history of apparent attacks on Hamilton's precedence in devising inclusive fitness theory, Hamilton's perspective on the issue, as well as a detailed analysis of the context in which Haldane presented his thought experiment, are to be found in Ullica Segerstrale's biography of Hamilton [Segerstrale, 2013].

2. "Haldane's dilemma" is a nod to the famous prisoner's dilemma of game theory, discussed in chapter 2, and differs from G. C. Williams's usage of the phrase as a chapter heading, which was in the context of the apparent logical constraints on what natural selection can achieve as a maximizing process [Williams, 1992].

3. Sterility has the same evolutionary outcome, in terms of direct fitness, as death. Contrary to the oft-repeated summary of Darwinism, coined by Herbert Spencer [Spencer, 1864], what matters for natural selection is not survival but reproduction.

4. To derive condition (9.3) we substitute the expressions for the direct fitness of the potential altruist (9.1) and the potential recipient (9.2) into the quantitative genetic equation for change in response to selection acting on inclusive fitness (8.4), then ask when the response to selection is positive (i.e., the right-hand side of (8.4) is positive). Since the individual in the river has no opportunity to exhibit any rescuing behavior, as we assume that potential drowning incidents are so rare that an individual will never experience more than one in their lifetime, whether as potential rescuer or potential rescuee, then the independent covariances are actually simple covariances; if we assume that rescuing behavior is unconditionally expressed and controlled by a single haploid gene with population frequency g, then these can be calculated as

$$\text{Cov}(W, P) = \text{E}(WP) - \text{E}(W)\text{E}(P)$$
$$= g\mu V - gV(1 - g(1 - \mu)) \tag{9.9}$$
$$= g(1 - g)V(\mu - 1)$$

and, similarly,

$$\text{Cov}(W', P) = g(1 - g)V(1 - \mu'). \tag{9.10}$$

Now, simply taking the right-hand side of condition (8.4), dividing through by narrow-sense heritability β_{GP} and by $\text{Var}(P) = g(1 - g)$, then asking when it is positive, we can rearrange to get

$$\frac{\beta_{G'P}}{\beta_{GP}} > \frac{V(1 - \mu)}{V(1 - \mu')}, \tag{9.11}$$

which can be simplified to give condition (9.3) in the main text.

5. The costs and benefits from this Hamilton's-rule analysis actually involve the values of the potential altruist's life and the potential recipient's life, as discussed in note 4.

6. See [Hamilton, 1964b] for very similar remarks.

7. The method used by Steve Frank to derive the extended version of Hamilton's rule (9.6) is based on maximization approaches and has not been described in this book, hence the derivation is not repeated here. See chapter 10, however, for a brief description of maximization approaches to building inclusive fitness models.

8. When $a = 1$, condition (9.6) simplifies to the condition

$$-c(1 - r) > 0, \tag{9.12}$$

which cannot be satisfied, since r cannot exceed 1, and is much less in most situations, such as in viscous populations.

9. The derivation of geometric compensated relatedness that follows is inspired by [Queller, 1994]. Starting with the observation that Hamilton's rule with compensated relatedness (9.8) describes all the fitness effects arising from altruism in an inelastic population, we then derive compensated relatedness in terms of the geometric genetic relatedness between potential donor and recipient (from HR4 in chapter 7), and the geometric relatedness between potential donor and the average individual from the potential recipient's neighborhood competing with them for reproduction, as

$$r_c = r - r_e$$
$$= \frac{\sum_P (G' - E(G)) P}{\sum_P (G - E(G)) P} - \frac{\sum_P (E(G^*) - E(G)) P}{\sum_P (G - E(G)) P} \tag{9.13}$$
$$= \frac{\sum_P (G' - E(G^*)) P}{\sum_P (G - E(G)) P},$$

where G^* is the genetic value of an individual facing increased competition from the recipient of an altruistic act. Compensated relatedness can be thus be interpreted in geometric terms (cf. figure 7.2), and from its geometric definition we can see that when the expected genetic value $E(G^*)$ of those experiencing increased competition due to an altruistic act equals the population mean genetic value $E(G)$ (competition is global), then compensated relatedness equals standard genetic relatedness. However, when the expected genetic value of competitors equals the genetic value of the potential recipient $(E(G^*) = G')$, competition is purely local, compensated relatedness is 0, and hence altruism cannot evolve.

10. As one would expect, the state-dependent strategy and individual state-dependent decisions that optimize fitness correspond to state-dependent versions of Hamilton's rule [McNamara et al. 1994a].

CHAPTER 10. EVIDENCE, OTHER APPROACHES, AND FURTHER TOPICS

1. In their *Nature* article, Martin Nowak, Corina Tarnita, and E. O. Wilson wrote,

> Considering its position for four decades as the dominant paradigm in the theoretical study of eusociality, the production of inclusive fitness theory must be considered meagre. [Nowak et al., 2010]

2. Note, however, that inclusive fitness is heritable in a way that group-related fitnesses are not, as discussed in chapter 8.

3. Note that group augmentation can also lead to indirect fitness benefits, leading to higher levels of helping than would be predicted by inclusive fitness considerations alone [Kokko et al., 2001].

4. Diploid eusocial species also exist, most notably the termites, but also including social shrimps, thrips, and aphids; see representative references within [Alpedrinha et al., 2013].

5. Interestingly, Hamilton's original thinking briefly anticipated a version of such results, when in discussing *Polistes* he wrote,

> There is good reason to believe that the initial nest-founding company is usually composed of sisters.... But it is doubtful if the wasps have any personal recognition of their sisters and if a wasp did arrive from far away it is probable that it would be accepted by the company provided it showed submission to the one or two highest ranking wasps. Dominance order does sometimes change and an accepted stranger has before it the prospect of rising in rank and ultimately subduing or driving off the queen. Thus an innocent rendering of assistance is not always easy to distinguish from an attempt at usurpations as Rau has pointed out, and the readiness to accept "help" is really just as puzzling as the disinterested assistance which some of the auxiliaries undoubtedly do render. [Hamilton, 1964b]

6. When the sex ratio is 50:50 then average relatedness between females and their brothers and sisters is $1/2 \times 1/4 + 1/2 \times 3/4 = 1/2$ [Trivers and Hare, 1976].

7. See references within [Alpedrinha et al., 2013].

8. On the status of the haplodiploidy hypothesis in terms of inclusive fitness theory see, for example, [Bourke, 2011b]. For a discussion of Hamilton's thinking on the issue see also [Segerstrale, 2013].

9. More recent proposals for the role of haplodiploidy in the evolution of eusociality include the promotion of maternal care [Linksvayer and Wade, 2005] (but see [Gardner, 2012]), promotion of helping through synergistic effects [Fromhage and Kokko, 2011], and facilitation of sex-ratio adjustment [Gardner and Ross, 2013].

10. In applying the monogamy hypothesis to the evolution of eusociality in haplodiploid species, a further assumption is needed of a *Fisherian sex ratio*, which is 50% male and 50% female, since then the average relatedness of an individual to their siblings is $1/2$ (see note 6). While queens favor an equal sex ratio, workers may attempt to achieve a female-biased sex ratio [Trivers and Hare, 1976]. Andrew Bourke and Nigel Franks, however, present a model predicting that haplodiploid workers should value siblings as much as offspring, regardless of the colony sex ratio ([Bourke and Franks, 1995, pp. 89–90]).

11. Interestingly, Hamilton's original papers also anticipated the development of the monogamy hypothesis. While noted for introducing the haplodiploidy hypothesis, Hamilton also discussed the potential route to eusociality in the diploid termites, in the context of an earlier evolutionary model of eusociality by George and Doris Williams [Williams and Williams, 1957], writing,

> The special considerations which apply to the *Hymenoptera* do not seem to have been noticed by Williams and Williams. The discussion which they base on their analysis of the full-sib relationship would, however, be applicable to the termites where this relationship is ensured in the colony by having the queen attended by a single "king." Termites of both sexes have an equal relationship ($r = 1/2$) to their siblings and their potential offspring. Thus the fact that both sexes "work" is just what we expect; we need only a bio-economic argument to explain why restriction of fertility to a few members has proved advantageous to the sibship as a whole. On this point the present theory can add little to previous discussions. [Hamilton, 1964b]

This is tantalisingly close to the monogamy hypothesis; the "bioeconomic" argument required by Hamilton, as provided by Boomsma [Boomsma, 2007, Boomsma, 2009], is that in some circumstances it becomes easier to raise siblings than directly reproduce, although this argument explains the benefits of foregoing direct reproduction in terms of inclusive fitness of individuals, rather than the efficiency of the whole colony.

12. Interestingly, as Boomsma points out [Boomsma, 2007], in introducing the haplodiploidy hypothesis Hamilton also noted the potential for multiple mating to arise once workers had lost their capacity to mate.

13. Martin Nowak and E. O. Wilson criticize the monogamy hypothesis by noting that only few monogamous species evolved to full eusociality [Wilson and Nowak, 2014]. This criticism misunderstands the necessary condition of monogamy as being a sufficient condition.

14. Unlike eusocial species in which, once workers have lost the ability to mate, promiscuity may evolve, in cooperatively breeding species helpers maintain the ability to mate, so if promiscuity by dominant breeders becomes the norm there is selective pressure against helping [Cornwallis et al., 2010].

15. Obligate multicellularity indicates that a species requires a multicellular stage in order for reproduction to occur. Individuals of facultatively multicellular species, such as *Dictyostelium* sp., can either reproduce via a multicellular route, as in slug, stalk, and fruiting body formation (figure 1.4), or reproduce independently [Fisher et al., 2013].

16. Hamilton's realization regarding the possibility for negative relatedness arose through his interactions with George Price and the regression coefficient formulation of genetic relatedness, as discussed in [Frank, 1995, Foster et al., 2001] and by Hamilton himself in [Hamilton, 1996].

17. E. O. Wilson proposed the interpretation of spite to negatively related individuals in terms of benefits to positively related individuals arising from reduced competition [Wilson, 1975b]. Laurent Lehman, Katja Bargum, and Max Reuter formally showed that Wilson's presentation of altruism as the flip side of spite is correct [Lehmann et al., 2006]. This result can be easily understood in terms of David Queller's reformulation of relatedness to take account of the competitive neighborhood of the recipients of social behavior (see chapter 9, note 9) [Gardner et al., 2007a]; by inspecting the version of Hamilton's rule using compensated relatedness (9.7) we can see that we can reinterpret a spiteful behavior $b < 0$ as an altruistic one by changing its sign and then also changing the sign of the compensated relatedness to leave the total inclusive fitness effect unchanged. Hence for the first term from condition (9.7) we have

$$b(r - r_e) = -b(r_e - r). \tag{10.1}$$

When we apply the geometric view of relatedness to calculate this new, negative compensated relatedness, as in chapter 9, note 9, we have

$$r_e - r = \frac{\sum_P (\mathrm{E}(G^*) - \mathrm{E}(G))P}{\sum_P (G - \mathrm{E}(G))P} - \frac{\sum_P (G' - \mathrm{E}(G))P}{\sum_P (G - \mathrm{E}(G))P}$$

$$= \frac{\sum_P (\mathrm{E}(G^*) - G')P}{\sum_P (G - \mathrm{E}(G))P}. \tag{10.2}$$

This new relatedness is the relatedness between actor (G) and individuals affected by competition changes resulting from the social behavior (G^*), calculated relative to the individuals directly affected by the social behavior (G'). So, a spiteful behavior that harms distantly related individuals can be interpreted as a behavior that benefits more closely related individuals, and conversely an altruistic behavior that benefits more closely related individuals can be interpreted as a behavior that harms more distantly related individuals. Note that this latter interpretation holds even when the competitive neighborhood of an individual is the general population ($\mathrm{E}(G^*) = \mathrm{E}(G)$) since in this case relatedness to the general population, calculated relative to the individuals helped as in condition (10.2), is still negative.

18. In fact, the Price equation is not an inherently nondynamic tool; rather, whether the Price equation can be applied recursively to describe long-term evolutionary dynamics is a property of the model of fitness one substitutes into it, as discussed in chapter 3 and in [Gardner et al., 2007b].

19. See note 18.

20. Among the genetic details that population genetical models can keep track of are the probabilities of alleles at different genetic loci co-occurring on the same genome. Satisfyingly, as Andy Gardner has shown together with Stuart West and Nicholas Barton, the multilocus approach to population genetics is analogous to Hamilton's inclusive fitness theory [Gardner et al., 2007b], which makes similar calculations but considers co-occurrence of alleles in different individuals.

21. Rousset's book also makes connections with methodologies including game theory and adaptive dynamics [Rousset, 2004].

22. The predictors used by Taylor and Frank are genetic value, used to predict own and partners' phenotypes [Taylor and Frank, 1996]. Thus the assumption of small variance in genetic value is an assumption of mutations of small effect, also presented as an assumption of small fitness effects and hence of small population fitness variance and, via Fisher's fundamental theorem, of weak selection.

23. Arne Traulsen, for example, contrasts inclusive fitness theory with evolutionary game theory [Traulsen, 2010]. In this book, however, we have used evolutionary game theory models to study selection acting on inclusive fitness; the former is a tool while the latter is a concept.

24. Hamilton's personal correspondence while conceiving inclusive fitness theory, indicating his hopes for it, is partly reproduced in [Segerstrale, 2013].

BIBLIOGRAPHY

[Abbot et al., 2011] Abbot, P., Abe, J., Alcock, J., Alizon, S., Alpedrinha, J. A. C., Andersson, M., Andre, J.-B., et al. (2011). Inclusive fitness theory and eusociality. *Nature*, 471:E1–E4.

[Allen et al., 2013] Allen, B., Nowak, M. A., and Wilson, E. O. (2013). Limitations of inclusive fitness. *Proceedings of the National Academy of Sciences*, 110:20135–20139.

[Alpedrinha et al., 2014] Alpedrinha, J., Gardner, A., and West, S. A. (2014). Haplodiploidy and the evolution of eusociality: worker revolution. *American Naturalist*, 184:303–317.

[Alpedrinha et al., 2013] Alpedrinha, J., West, S. A., and Gardner, A. (2013). Haplodiploidy and the evolution of eusociality: worker reproduction. *American Naturalist*, 182(4):421–438.

[Archetti and Scheuring, 2011] Archetti, M. and Scheuring, I. (2011). Coexistence of cooperation and defection in public goods games. *Evolution*, 65(4):1140–1148.

[Archetti and Scheuring, 2012] Archetti, M. and Scheuring, I. (2012). Review: game theory of public goods in one-shot social dilemmas without assortment. *Journal of Theoretical Biology*, 299:9–20.

[Axelrod, 1984] Axelrod, R. (1984). *The Evolution of Cooperation*. Basic Books.

[Axelrod and Hamilton, 1981] Axelrod, R. and Hamilton, W. D. (1981). The evolution of cooperation. *Science*, 211(4489):1390–1396.

[Bateson, 1909] Bateson, W. (1909). *Mendel's Principles of Heredity*. Cambridge University Press, Cambridge.

[Biernaskie et al., 2011] Biernaskie, J. M., West, S. A., and Gardner, A. (2011). Are greenbeards intragenomic outlaws? *Evolution*, 65(10):2729–2742.

[Birch, 2013] Birch, J. (2013). *Kin Selection: A Philosophical Analysis*. PhD thesis, University of Cambridge.

[Birch and Marshall, 2014] Birch, J. and Marshall, J. A. R. (2014). Queller's separation condition explained and defended. *American Naturalist* 184, 531–540.

[Boomsma, 2007] Boomsma, J. J. (2007). Kin selection versus sexual selection: why the ends do not meet. *Current Biology*, 17(16):R673–R683.

[Boomsma, 2009] Boomsma, J. J. (2009). Lifetime monogamy and the evolution of eusociality. *Philosophical Transactions of the Royal Society B: Biological Sciences*, 364(1533):3191–3207.

[Boomsma, 2013] Boomsma, J. J. (2013). Beyond promiscuity: mate-choice commitments in social breeding. *Philosophical Transactions of the Royal Society B: Biological Sciences*, 368.

[Boomsma et al., 2011] Boomsma, J. J., Beekman, M., Cornwallis, C. K., Griffin, A. S., Holman, L., Hughes, W. O., Keller, L., Oldroyd, B. P., and Ratnieks, F. (2011). Only full-sibling families evolved eusociality. *Nature*, 471:E4–E5.

[Boorman and Levitt, 1980] Boorman, S. A. and Levitt, P. R. (1980). *The Genetics of Altruism*. Academic Press.

[Bourke, 2011a] Bourke, A. F. G. (2011a). *Principles of Social Evolution*. Oxford University Press, Oxford.

[Bourke, 2011b] Bourke, A. F. G. (2011b). The validity and value of inclusive fitness theory. *Proceedings of the Royal Society B: Biological Sciences*, 278(1723):3313–3320.

[Bourke, 2014] Bourke, A. F. G. (2014). Hamilton's rule and the causes of social evolution. *Philosophical Transactions of the Royal Society B: Biological Sciences*, 369. doi:10.1098/rstb.2013.0362.

[Bourke and Franks, 1995] Bourke, A. F. G. and Franks, N. R. (1995). *Social Evolution in Ants*. Princeton University Press.

[Boyd et al., 2011] Boyd, R., Richerson, P. J., and Henrich, J. (2011). Rapid cultural adaptation can facilitate the evolution of large-scale cooperation. *Behavioral Ecology and Sociobiology*, 65(3): 431–444.

[Burda et al., 2000] Burda, H., Honeycutt, R. L., Begall, S., Locker-Grütjen, O., and Scharff, A. (2000). Are naked and common mole-rats eusocial and if so, why? *Behavioral Ecology and Sociobiology*, 47(5):293–303.

[Buss, 1987] Buss, L. (1987). *The Evolution of Individuality*. Princeton University Press.

[Cavalli-Sforza and Feldman, 1978] Cavalli-Sforza, L. L. and Feldman, M. W. (1978). Darwinian selection and "altruism". *Theoretical Population Biology*, 14(2):268–280.

[Chittka and Chittka, 2010] Chittka, A. and Chittka, L. (2010). Epigenetics of royalty. *PLoS Biology*, 8(11):e1000532.

[Clutton-Brock, 2002] Clutton-Brock, T. (2002). Breeding together: kin selection and mutualism in cooperative vertebrates. *Science*, 296(5565):69–72.

[Clutton-Brock, 2009] Clutton-Brock, T. (2009). Cooperation between non-kin in animal societies. *Nature*, 462(7269):51–57.

[Clutton-Brock et al., 2001] Clutton-Brock, T. H., Brotherton, P. N. M., O'Riain, M. J., Griffin, A. S., Gaynor, D., Kansky, R., Sharpe, L., and McIlrath, G. M. (2001). Contributions to cooperative rearing in meerkats. *Animal Behaviour*, 61(4):705–710.

[Clutton-Brock et al., 2000] Clutton-Brock, T. H., Brotherton, P. N. M., O'Riain, M. J., Griffin, A. S., Gaynor, D., Sharpe, L., Kansky, R., Manser, M. B., and McIlrath, G. M. (2000). Individual contributions to babysitting in a cooperative mongoose, *Suricata suricatta*. *Proceedings of the Royal Society B: Biological Sciences*, 267(1440):301–305.

[Clutton-Brock et al., 1998] Clutton-Brock, T. H., Gaynor, D., Kansky, R., MacColl, A. D. C., McIlrath, G., Chadwick, P., Brotherton, P. N. M., O'Riain, M. J., Manser, M., and Skinner, J. D. (1998). Costs of cooperative behaviour in suricates (*Suricata suricatta*). *Proceedings of the Royal Society B: Biological Sciences*, 265(1392):185–190.

[Clutton-Brock et al., 1999] Clutton-Brock, T. H., O'Riain, M. J., Brotherton, P. N. M., Gaynor, D., Kansky, R., Griffin, A. S., and Manser, M. (1999). Selfish sentinels in cooperative mammals. *Science*, 284(5420):1640–1644.

[Cornwallis et al., 2010] Cornwallis, C. K., West, S. A., Davis, K. E., and Griffin, A. S. (2010). Promiscuity and the evolutionary transition to complex societies. *Nature*, 466(7309):969–972.

[Craig, 1979] Craig, R. (1979). Parental manipulation, kin selection, and the evolution of altruism. *Evolution*, 33:319–334.

[Crespi, 2001] Crespi, B. J. (2001). The evolution of social behavior in microorganisms. *Trends in Ecology & Evolution*, 16(4):178–183.

[Crespi and Yanega, 1995] Crespi, B. J. and Yanega, D. (1995). The definition of eusociality. *Behavioral Ecology*, 6(1):109–115.

[Crozier, 1986] Crozier, R. H. (1986). Genetic clonal recognition abilities in marine invertebrates must be maintained by selection for something else. *Evolution*, 40(5):1100–1101.

[Darwin, 1859] Darwin, C. (1859). *On the Origin of Species by Means of Natural Selection, or the Preservation of Favoured Races in the Struggle for Life*. John Murray, 6th edition.

[Darwin, 1871] Darwin, C. (1871). *The Descent of Man, and Selection in Relation to Sex*. John Murray, London.

[Darwin and Wallace, 1858] Darwin, C. and Wallace, A. R. (1858). On the tendency of species to form varieties; and on the perpetuation of varieties and species by means of natural selection. *Journal of the Proceedings of the Linnean Society: Zoology*, 3:45–62.

[Dawkins, 1976] Dawkins, R. (1976). *The Selfish Gene*. Oxford University Press.

[Day and Taylor, 1997] Day, T. and Taylor, P. D. (1997). Hamilton's rule meets the Hamiltonian: kin selection on dynamic characters. *Proceedings of the Royal Society B: Biological Sciences*, 264:639–644.

[Dieckmann, 1985] Dieckmann, A. (1985). Volunteer's dilemma. *Journal of Conflict Resolution*, 29:605–610.

[Dijkstra and Boomsma, 2006] Dijkstra, M. B. and Boomsma, J. J. (2006). Are workers of *Atta* leafcutter ants capable of reproduction? *Insectes Sociaux*, 53(2):136–140.

[Dijkstra et al., 2005] Dijkstra, M. B., Nash, D. R., and Boomsma, J. J. (2005). Self-restraint and sterility in workers of *Acromyrmex* and *Atta* leafcutter ants. *Insectes Sociaux*, 52(1):67–76.

[Edwards, 2013] Edwards, A. W. F. (2013). R.A. Fisher's gene-centred view of evolution and the Fundamental Theorem of Natural Selection. *Biological Reviews*, 89:135–147.

[Emlen, 1982] Emlen, S. T. (1982). The evolution of helping. II. The role of behavioral conflict. *American Naturalist*, 119:40–53.

[Falconer and Mackay, 1996] Falconer, D. S. and Mackay, T. F. C. (1996). *Introduction to Quantitative Genetics*. Pearson, fourth edition.

[Ferriere and Michod, 2011] Ferriere, R. and Michod, R. E. (2011). Inclusive fitness in evolution. *Nature*, 471:E6–E8.

[Fisher, 1930] Fisher, R. A. (1930). *The Genetical Theory of Natural Selection*. Clarendon Press, Oxford.

[Fisher, 1941] Fisher, R. A. (1941). Average excess and average effect of a gene substitution. *Annals of Eugenics*, 11(1):53–63.

[Fisher et al., 2013] Fisher, R. M., Cornwallis, C. K., and West, S. A. (2013). Group formation, relatedness, and the evolution of multicellularity. *Current Biology*, 23:1120–1125.

[Fletcher and Doebeli, 2009] Fletcher, J. A. and Doebeli, M. (2009). A simple and general explanation for the evolution of altruism. *Proceedings of the Royal Society B: Biological Sciences*, 276:13–19.

[Fletcher and Zwick, 2006] Fletcher, J. A. and Zwick, M. (2006). Unifying the theories of inclusive fitness and reciprocal altruism. *American Naturalist*, 168(2):252–262.

[Fletcher et al., 2006] Fletcher, J. A., Zwick, M., Doebeli, M., and Wilson, D. S. (2006). What's wrong with inclusive fitness? *Trends in Ecology and Evolution*, 21(11):597–598.

[Foster, 2008] Foster, K. R. (2008). Behavioral ecology: altruism. *Encyclopedia of Ecology*, pages 154–159.

[Foster, 2009] Foster, K. R. (2009). A defense of sociobiology. In *Cold Spring Harbor Symposia on Quantitative Biology*, volume 74, pages 403–418. Cold Spring Harbor Laboratory Press.

[Foster, 2011] Foster, K. R. (2011). The secret social lives of microorganisms. *Microbe*, 6:183–186.

[Foster and Wenseleers, 2006] Foster, K. R. and Wenseleers, T. (2006). A general model for the evolution of mutualisms. *Journal of Evolutionary Biology*, 19(4):1283–1293.

[Foster et al., 2001] Foster, K. R., Wenseleers, T., and Ratnieks, F. L. W. (2001). Spite: Hamilton's unproven theory. *Annales Zoologici Fennici*, 38:229–238.

[Foster et al., 2006a] Foster, K. R., Wenseleers, T., and Ratnieks, F. L. W. (2006a). Kin selection is the key to altruism. *Trends in Ecology and Evolution*, 21(2):57–60.

[Foster et al., 2006b] Foster, K. R., Wenseleers, T., Ratnieks, F. L. W., and Queller, D. C. (2006b). There is nothing wrong with inclusive fitness. *Trends in Ecology and Evolution*, 21(11):599–600.

[Frank, 1994] Frank, S. A. (1994). Genetics of mutualism: the evolution of altruism between species. *Journal of Theoretical Biology*, 170(4):393–400.

[Frank, 1995] Frank, S. A. (1995). George Price's contributions to evolutionary genetics. *Journal of Theoretical Biology*, 175(3):373–388.

[Frank, 1997] Frank, S. A. (1997). The Price equation, Fisher's fundamental theorem, kin selection, and causal analysis. *Evolution*, 51:1712–1729.

[Frank, 1998] Frank, S. A. (1998). *Foundations of Social Evolution*. Princeton University Press.

[Frank, 2012] Frank, S. A. (2012). Natural selection. IV. The Price equation. *Journal of Evolutionary Biology*, 25(6):1002–1019.

[Frank, 2013] Frank, S. A. (2013). Natural selection. VII. History and interpretation of kin selection theory. *Journal of Evolutionary Biology*, 26(6):1151–1184.

[Frank and Slatkin, 1992] Frank, S. A. and Slatkin, M. (1992). Fisher's fundamental theorem of natural selection. *Trends in Ecology and Evolution*, 7(3):92–95.

[Fromhage and Kokko, 2011] Fromhage, L. and Kokko, H. (2011). Monogamy and haplodiploidy act in synergy to promote the evolution of eusociality. *Nature Communications*, 2:397.

[Futuyma, 1998] Futuyma, D. J. (1998). *Evolutionary Biology*. Sinauer, third edition.

[Gardner, 2008] Gardner, A. (2008). The Price equation. *Current Biology*, 18(5):R198–R202.

[Gardner, 2009] Gardner, A. (2009). Adaptation as organism design. *Biology Letters*, 5(6):861–864.

[Gardner, 2012] Gardner, A. (2012). Evolution of maternal care in diploid and haplodiploid populations. *Journal of Evolutionary Biology*, 25:1479–1486.

[Gardner et al., 2012] Gardner, A., Alpedrinha, J., and West, S. A. (2012). Haplodiploidy and the evolution of eusociality: split sex ratios. *American Naturalist*, 179(2):240–256.

[Gardner and Foster, 2008] Gardner, A. and Foster, K. R. (2008). The evolution and ecology of cooperation–history and concepts. In *Ecology of Social Evolution*, pages 1–36. Springer.

[Gardner and Grafen, 2009] Gardner, A. and Grafen, A. (2009). Capturing the superorganism: a formal theory of group adaptation. *Journal of Evolutionary Biology*, 22:659–671.

[Gardner et al., 2007a] Gardner, A., Hardy, I. C. W., Taylor, P. D., and West, S. A. (2007a). Spiteful soldiers and sex ratio conflict in polyembryonic parasitoid wasps. *American Naturalist*, 169(4):519–533.

[Gardner and Ross, 2013] Gardner, A. and Ross, L. (2013). Haplodiploidy, sex-ratio adjustment, and eusociality. *American Naturalist*, 181:E60–E67.

[Gardner and Welch, 2011] Gardner, A. and Welch, J. J. (2011). A formal theory of the selfish gene. *Journal of Evolutionary Biology*, 24:1801–1813.

[Gardner and West, 2007] Gardner, A. and West, S. A. (2007). Social evolution: the decline and fall of genetic kin recognition. *Current Biology*, 17(18):R810–R812.

[Gardner and West, 2010] Gardner, A. and West, S. A. (2010). Greenbeards. *Evolution*, 64(1):25–38.

[Gardner et al., 2007b] Gardner, A., West, S. A., and Barton, N. H. (2007b). The relation between multilocus population genetics and social evolution theory. *American Naturalist*, 169(2):207–226.

[Gardner et al., 2011] Gardner, A., West, S. A., and Wild, G. (2011). The genetical theory of kin selection. *Journal of Evolutionary Biology*, 24(5):1020–1043.

[Gilbert et al., 2007] Gilbert, O. M., Foster, K. R., Mehdiabadi, N. J., Strassmann, J. E., and Queller, D. C. (2007). High relatedness maintains multicellular cooperation in a social amoeba by controlling cheater mutants. *Proceedings of the National Academy of Sciences*, 104:8913–8917.

[Grafen, 1979] Grafen, A. (1979). The hawk–dove game played between relatives. *Animal Behaviour*, 27:905–907.

[Grafen, 1984] Grafen, A. (1984). *Natural Selection, Kin Selection and Group Selection*, volume Behavioural Ecology: An Evolutionary Approach, chapter 3. Blackwell Publishing, 2nd edition.

[Grafen, 1985a] Grafen, A. (1985a). A geometric view of relatedness. *Oxford Surveys in Evolutionary Biology*, 2:28–89.

[Grafen, 1985b] Grafen, A. (1985b). Hamilton's rule OK. *Nature*, 318:310–311.

[Grafen, 2000] Grafen, A. (2000). Developments of the Price equation and natural selection under uncertainty. *Proceedings of the Royal Society B: Biological Sciences*, 267(1449):1223–1227.

[Grafen, 2006a] Grafen, A. (2006a). Optimization of inclusive fitness. *Journal of Theoretical Biology*, 238(3):541–563.

[Grafen, 2006b] Grafen, A. (2006b). Various remarks on Lehmann and Keller's article. *Journal of Evolutionary Biology*, 19(5):1397–1399.

[Grafen, 2007a] Grafen, A. (2007a). The formal Darwinism project: a mid-term report. *Journal of Evolutionary Biology*, 20(4):1243–1254.

[Grafen, 2007b] Grafen, A. (2007b). An inclusive fitness analysis of altruism on a cyclical network. *Journal of Evolutionary Biology*, 20(6):2278–2283.

[Grafen, 2014a] Grafen, A. (2014a). The formal Darwinism project in outline. *Biology and Philosophy*, 29:155–174.

[Grafen, 2014b] Grafen, A. (2014b). The formal Darwinism project in outline: response to commentaries. *Biology and Philosophy*, 29:281–292.

[Grafen and Archetti, 2008] Grafen, A. and Archetti, M. (2008). Natural selection of altruism in inelastic viscous homogeneous populations. *Journal of Theoretical Biology*, 252(4):694–710.

[Grafen and Hails, 2002] Grafen, A. and Hails, R. (2002). *Modern Statistics for the Life Sciences*. Oxford University Press.

[Griffin and West, 2002] Griffin, A. S. and West, S. A. (2002). Kin selection: fact and fiction. *Trends in Ecology & Evolution*, 17(1):15–21.

[Griffin and West, 2003] Griffin, A. S. and West, S. A. (2003). Kin discrimination and the benefit of helping in cooperatively breeding vertebrates. *Science*, 302:634–636.

[Griffin et al., 2004] Griffin, A. S., West, S. A., and Buckling, A. (2004). Cooperation and competition in pathogenic bacteria. *Nature*, 430(7003):1024–1027.

[Hadfield et al., 2010] Hadfield, J. D., Wilson, A. J., Garant, D., Sheldon, B. C., and Kruuk, L. E. B. (2010). The misuse of blup in ecology and evolution. *American Naturalist*, 175(1):116–125.

[Haldane, 1932] Haldane, J. B. S. (1932). *The Causes of Evolution*. Longmans, London.

[Haldane, 1955] Haldane, J. B. S. (1955). Population genetics. *New Biology*, 18:34–51.

[Hamilton, 1963] Hamilton, W. D. (1963). The evolution of altruistic behavior. *American Naturalist*, 97(896):354–356.

[Hamilton, 1964a] Hamilton, W. D. (1964a). The genetical evolution of social behaviour I. *Journal of Theoretical Biology*, 7:1–16.

[Hamilton, 1964b] Hamilton, W. D. (1964b). The genetical evolution of social behaviour II. *Journal of Theoretical Biology*, 7:17–52.

[Hamilton, 1967] Hamilton, W. D. (1967). Extraordinary sex ratios. *Science*, 156(3774):477–488.

[Hamilton, 1970] Hamilton, W. D. (1970). Selfish and spiteful behaviour in an evolutionary model. *Nature*, 228(5277):1218–1220.

[Hamilton, 1971] Hamilton, W. D. (1971). Selection of selfish and altruistic behavior in some extreme models. In *Man and Beast: Comparative Social Behavior*, pages 57–91. Smithsonian Press.

[Hamilton, 1972] Hamilton, W. D. (1972). Altruism and related phenomena, mainly in social insects. *Annual Review of Ecology and Systematics*, 3(1):193–232.

[Hamilton, 1975] Hamilton, W. D. (1975). Innate social aptitudes of man: an approach from evolutionary genetics. *Biosocial Anthropology*, pages 133–155.

[Hamilton, 1996] Hamilton, W. D. (1996). *Narrow Roads of Gene Land Volume 1: Evolution of Social Behaviour*. Oxford University Press.

[Hamilton, 2001] Hamilton, W. D. (2001). *Narrow Roads of Gene Land Volume 2: Evolution of Sex*. Oxford University Press.

[Hamilton, 2005] Hamilton, W. D. (2005). *Narrow Roads of Gene Land Volume 3: Last Words*. Oxford University Press.

[Harman, 2010] Harman, O. S. (2010). *The Price of Altruism: George Price and the Search for the Origins of Kindness*. Bodley Head.

[Hatchwell, 2009] Hatchwell, B. J. (2009). The evolution of cooperative breeding in birds: kinship, dispersal and life history. *Philosophical Transactions of the Royal Society B: Biological Sciences*, 364:3217–3227.

[Hatchwell et al., 2014] Hatchwell, B. J., Gullett, P. R., and Adams, M. J. (2014). Helping in cooperatively breeding long-tailed tits: a test of Hamilton's rule. *Philosophical Transactions of the Royal Society B: Biological Sciences*, 369.

[Hatchwell et al., 2004] Hatchwell, B. J., Russell, A. F., MacColl, A. D. C., Ross, D. J., Fowlie, M. K., and McGowan, A. (2004). Helpers increase long-term but not short-term productivity in cooperatively breeding long-tailed tits. *Behavioral Ecology*, 15(1):1–10.

[Hatchwell and Sharp, 2006] Hatchwell, B. J. and Sharp, S. P. (2006). Kin selection, constraints, and the evolution of cooperative breeding in long-tailed tits. *Advances in the Study of Behavior*, 36: 355–395.

[Hauert et al., 2006] Hauert, C., Michor, F., Nowak, M. A., and Doebeli, M. (2006). Synergy and discounting of cooperation in social dilemmas. *Journal of Theoretical Biology*, 239(2):195–202.

[Heisler and Damuth, 1987] Heisler, I. L. and Damuth, J. (1987). A method for analyzing selection in hierarchically structured populations. *American Naturalist*, 130:582–602.

[Herre and Wcislo, 2011] Herre, E. A. and Wcislo, W. T. (2011). In defence of inclusive fitness theory. *Nature*, 471:E8–E9.

[Houston and McNamara, 1999] Houston, A. I. and McNamara, J. M. (1999). *Models of Adaptive Behaviour: An Approach Based on State*. Cambridge University Press.

[Hughes et al., 2008] Hughes, W. O. H., Oldroyd, B. P., Beekman, M., and Ratnieks, F. L. W. (2008). Ancestral monogamy shows kin selection is key to the evolution of eusociality. *Science*, 320(5880):1213–1216.

[Huxley, 1942] Huxley, J. S. (1942). *Evolution: The Modern Synthesis*. Unwin, London.

[Jarvis, 1981] Jarvis, J. U. (1981). Eusociality in a mammal: cooperative breeding in naked mole-rat colonies. *Science*, 212(4494):571–573.

[Jiricny et al., 2010] Jiricny, N., Diggle, S. P., West, S. A., Evans, B. A., Ballantyne, G., Ross-Gillespie, A., and Griffin, A. S. (2010). Fitness correlates with the extent of cheating in a bacterium. *Journal of Evolutionary Biology*, 23(4):738–747.

[Kokko et al., 2001] Kokko, H., Johnstone, R. A., and Clutton-Brock, T. H. (2001). The evolution of cooperative breeding through group augmentation. *Proceedings of the Royal Society B: Biological Sciences*, 268(1463):187–196.

[Kruuk et al., 2014] Kruuk, L., Clutton-Brock, T., and Pemberton, J. (2014). Case study: quantitative genetics and sexual selection of weaponry in a wild ungulate. In Charmantier, A., Garant, D., and Kruuk, L. E. B., editors, *Quantitative Genetics in the Wild*, pages 160–176. Oxford University Press.

[Kruuk et al., 2002] Kruuk, L. E. B., Slate, J., Pemberton, J. M., Brotherstone, S., Guinness, F., and Clutton-Brock, T. (2002). Antler size in red deer: heritability and selection but no evolution. *Evolution*, 56(8):1683–1695.

[Lahdenperä et al., 2004] Lahdenperä, M., Lummaa, V., Helle, S., Tremblay, M., and Russell, A. F. (2004). Fitness benefits of prolonged post-reproductive lifespan in women. *Nature*, 428:178–181.

[Lande and Arnold, 1983] Lande, R. and Arnold, S. J. (1983). The measurement of selection on correlated characters. *Evolution*, 37:1210–1226.

[Leadbeater et al., 2011] Leadbeater, E., Carruthers, J. M., Green, J. P., Rosser, N. S., and Field, J. (2011). Nest inheritance is the missing source of direct fitness in a primitively eusocial insect. *Science*, 333(6044):874–876.

[Lehmann et al., 2006] Lehmann, L., Bargum, K., and Reuter, M. (2006). An evolutionary analysis of the relationship between spite and altruism. *Journal of Evolutionary Biology*, 19(5):1507–1516.

[Lehmann and Keller, 2006a] Lehmann, L. and Keller, L. (2006a). Synergy, partner choice and frequency dependence: their integration into inclusive fitness theory and their interpretation in terms of direct and indirect fitness effects. *Journal of Evolutionary Biology*, 19(5):1426–1436.

[Lehmann and Keller, 2006b] Lehmann, L. and Keller, L. (2006b). The evolution of cooperation and altruism – a general framework and a classification of models. *Journal of Evolutionary Biology*, 19(5):1365–1376.

[Lehmann et al., 2007] Lehmann, L., Keller, L., and Sumpter, D. J. T. (2007). The evolution of helping and harming on graphs: the return of the inclusive fitness effect. *Journal of Evolutionary Biology*, 20(6):2284–2295.

[Lehmann and Rousset, 2014] Lehmann, L. and Rousset, F. (2014). Fitness, inclusive fitness, and optimisation. *Biology and Philosophy*, 29:181–195.

[Lewontin, 1970] Lewontin, R. C. (1970). The units of selection. *Annual Review of Ecology and Systematics*, 1:1–18.

[Li, 1967]   Li, C. C. (1967). Fundamental theorem of natural selection. *Nature*, 214:505–506.

[Liao et al., 2015]   Liao, X., Rong, S., and Queller, D. (2015). Relatedness, conflict and the evolution of eusociality. *PLoS Biology* (in press).

[Linksvayer and Wade, 2005]   Linksvayer, T. A. and Wade, M. J. (2005). The evolutionary origin and elaboration of sociality in the aculeate *Hymenoptera*: maternal effects, sib-social effects, and heterochrony. *Quarterly Review of Biology*, 80:317–336.

[Lukas and Clutton-Brock, 2012]   Lukas, D. and Clutton-Brock, T. (2012). Cooperative breeding and monogamy in mammalian societies. *Proceedings of the Royal Society B: Biological Sciences*, 279(1736):2151–2156.

[MacLeod et al., 2013]   MacLeod, K. J., Nielsen, J. F., and Clutton-Brock, T. H. (2013). Factors predicting the frequency, likelihood and duration of allonursing in the cooperatively breeding meerkat. *Animal Behaviour*, 86:1059–1067.

[Marshall, 2009]   Marshall, J. A. R. (2009). The donation game with roles played between relatives. *Journal of Theoretical Biology*, 260(3):386–391.

[Marshall, 2011a]   Marshall, J. A. R. (2011a). Group selection and kin selection: formally equivalent approaches. *Trends in Ecology & Evolution*, 26(7):325–332.

[Marshall, 2011b]   Marshall, J. A. R. (2011b). Queller's rule ok: Comment on van Veelen "when inclusive fitness is right and when it can be wrong". *Journal of Theoretical Biology*, 270(1): 185–188.

[Marshall, 2011c]   Marshall, J. A. R. (2011c). Ultimate causes and the evolution of altruism. *Behavioral Ecology and Sociobiology*, 65(3):503–512.

[Marshall, 2014]   Marshall, J. A. R. (2014). Generalisations of Hamilton's rule applied to non-additive public goods games with random group size. *Frontiers in Ecology and Evolution*. doi:10.3389/fevo.2014.00040.

[Marshall and Rowe, 2003]   Marshall, J. A. R. and Rowe, J. E. (2003). Kin selection may inhibit the evolution of reciprocation. *Journal of Theoretical Biology*, 222(3):331–335.

[Maynard Smith, 1964]   Maynard Smith, J. (1964). Group selection and kin selection. *Nature*, 201:1145–1147.

[Maynard Smith, 1982]   Maynard Smith, J. (1982). *Evolution and the Theory of Games*. Cambridge University Press.

[Maynard Smith and Parker, 1976]   Maynard Smith, J. and Parker, G. A. (1976). The logic of asymmetric contests. *Animal Behaviour*, 24(1):159–175.

[Maynard Smith and Price, 1973]   Maynard Smith, J. and Price, G. R. (1973). The logic of animal conflict. *Nature*, 246:15–18.

[Maynard Smith and Szathmáry, 1997]   Maynard Smith, J. and Szathmáry, E. (1997). *The Major Transitions in Evolution*. Oxford University Press.

[McElreath and Boyd, 2008]   McElreath, R. and Boyd, R. (2008). *Mathematical Models of Social Evolution: A Guide for the Perplexed*. University of Chicago Press.

[McGlothlin et al., 2014]   McGlothlin, J., Wolf, J., Brodie III, E., and Moore, A. (2014). Quantitative genetic versions of Hamilton's rule with empirical applications. *Philosophical Transactions of the Royal Society B: Biological Sciences*, 369. doi:10.1098/rstb.2013.0358.

[McNamara and Houston, 1996]   McNamara, J. M. and Houston, A. I. (1996). State-dependent life histories. *Nature*, 380:215–221.

[McNamara et al., 1994]   McNamara, J. M., Houston, A. I., and Webb, J. N. (1994). Dynamic kin selection. *Proceedings of the Royal Society B: Biological Sciences*, 258:23–28.

[McNamara et al., 2011]   McNamara, J. M., Trimmer, P. C., Eriksson, A., Marshall, J. A. R., and Houston, A. I. (2011). Environmental variability can select for optimism or pessimism. *Ecology Letters*, 14(1):58–62.

[Meade and Hatchwell, 2010]   Meade, J. and Hatchwell, B. J. (2010). No direct fitness benefits of helping in a cooperative breeder despite higher survival of helpers. *Behavioral Ecology*, 21(6): 1186–1194.

[Mehdiabadi et al., 2006] Mehdiabadi, N. J., Jack, C. N., Farnham, T. T., Platt, T. G., Kalla, S. E., Shaulsky, G., Queller, D. C., and Strassmann, J. E. (2006). Social evolution: kin preference in a social microbe. *Nature*, 442(7105):881–882.

[Mendl, 1866] Mendl, G. (1866). Versuche über pflanzen-hybriden. *Verhandlungen des Naturforschenden Vereines in Brunn*, IV:3–47.

[Metz et al., 1992] Metz, J. A. J., Nisbet, R. M., and Geritz, S. A. H. (1992). How should we define "fitness" for general ecological scenarios? *Trends in Ecology & Evolution*, 7(6):198–202.

[Michel-Briand and Baysse, 2002] Michel-Briand, Y. and Baysse, C. (2002). The pyocins of *Pseudomonas aeruginosa*. *Biochimie*, 84(5):499–510.

[Michod, 2000] Michod, R. E. (2000). *Darwinian Dynamics: Evolutionary Transitions in Fitness and Individuality*. Princeton University Press.

[Michod and Hamilton, 1980] Michod, R. E. and Hamilton, W. D. (1980). Coefficients of relatedness in sociobiology. *Nature*, 288(5792):694–697.

[Morrissey et al., 2010] Morrissey, M. B., Kruuk, L. E. B., and Wilson, A. J. (2010). The danger of applying the breeder's equation in observational studies of natural populations. *Journal of Evolutionary Biology*, 23(11):2277–2288.

[Motro, 1991] Motro, U. (1991). Co-operation and defection: playing the field and the ess. *Journal of Theoretical Biology*, 151(2):145–154.

[Nam et al., 2010] Nam, K., Simeoni, M., Sharp, S. P., and Hatchwell, B. J. (2010). Kinship affects investment by helpers in a cooperatively breeding bird. *Proceedings of the Royal Society B: Biological Sciences*, 277(1698):3299–3306.

[Nasar, 1998] Nasar, S. (1998). *A Beautiful Mind: A Biography of John Forbes Nash, Jr., Winner of the Nobel Prize in Economics, 1994*. Simon and Schuster.

[Nash, 1951] Nash, J. (1951). Non-cooperative games. *Annals of Mathematics*, 54(2):286–295.

[Nash, 1950] Nash, J. F. (1950). Equilibrium points in n-person games. *Proceedings of the National Academy of Sciences*, 36(1):48–49.

[Nee, 1989] Nee, S. (1989). Does Hamilton's rule describe the evolution of reciprocal altruism? *Journal of Theoretical Biology*, 141(1):81–91.

[Nowak et al., 2010] Nowak, M. A., Tarnita, C. E., and Wilson, E. O. (2010). The evolution of eusociality. *Nature*, 466(7310):1057–1062.

[Nowak et al., 2011] Nowak, M. A., Tarnita, C. E., and Wilson, E. O. (2011). Nowak *et al.* reply. *Nature*, 471:E9–E10.

[Nunney, 1985] Nunney, L. (1985). Group selection, altruism, and structured-deme models. *American Naturalist*, 126:212–230.

[Ohtsuki et al., 2006] Ohtsuki, H., Hauert, C., Lieberman, E., and Nowak, M. A. (2006). A simple rule for the evolution of cooperation on graphs. *Nature*, 441(7092):502.

[Ohtsuki and Nowak, 2006] Ohtsuki, H. and Nowak, M. A. (2006). Evolutionary games on cycles. *Proceedings of the Royal Society B: Biological Sciences*, 273(1598):2249–2256.

[Okasha, 2006] Okasha, S. (2006). *Evolution and the Levels of Selection*. Oxford University Press.

[Okasha, 2008] Okasha, S. (2008). Fisher's fundamental theorem of natural selection—a philosophical analysis. *British Journal for the Philosophy of Science*, 59(3):319–351.

[Okasha, 2014] Okasha, S. (2014). The relation between kin and multi-level selection: an approach using causal graphs. *British Journal for the Philosophy of Science*. In press.

[Okasha and Paternotte, 2012] Okasha, S. and Paternotte, C. (2012). Group adaptation, formal Darwinism and contextual analysis. *Journal of Evolutionary Biology*, 25:1127–1139.

[Okasha and Paternotte, 2014] Okasha, S. and Paternotte, C. (2014). Adaptation, fitness and the selection-optimality links. *Biology and Philosophy*, 29:225–232.

[Oli, 2003] Oli, M. K. (2003). Hamilton goes empirical: estimation of inclusive fitness from life-history data. *Proceedings of the Royal Society B: Biological Sciences*, 270:307–311.

[Olson, 1965] Olson, M. (1965). *The Logic of Collective Action: Public Goods and the Theory of Groups*. Harvard University Press.

[Osinga and Marshall, 2015] Osinga, H. and Marshall, J. A. R. (2015). Adaptive topographies and equilibrium selection in an evolutionary game. *PLoS One* (in press).

[Paley, 1802] Paley, W. (1802). *Natural Theology: or, Evidence of the Existence and Attributes of the Deity*. Faulder, London.

[Platt and Bever, 2009] Platt, T. G. and Bever, J. D. (2009). Kin competition and the evolution of cooperation. *Trends in Ecology and Evolution*, 24:370–377.

[Price, 1970] Price, G. R. (1970). Selection and covariance. *Nature*, 227(5257):520–521.

[Price, 1972a] Price, G. R. (1972a). Extension of covariance selection mathematics. *Annals of Human Genetics*, 35:485–490.

[Price, 1972b] Price, G. R. (1972b). Fisher's "fundamental theorem" made clear. *Annals of Human Genetics*, 36(2):129–140.

[Queller, 1984] Queller, D. C. (1984). Kin selection and frequency dependence: A game theoretic approach. *Biological Journal of the Linnean Society*, 23(2):133–143.

[Queller, 1985] Queller, D. C. (1985). Kinship, reciprocity and synergism in the evolution of social behaviour. *Nature*, 318(6044):366–367.

[Queller, 1992a] Queller, D. C. (1992a). A general model for kin selection. *Evolution*, pages 376–380.

[Queller, 1992b] Queller, D. C. (1992b). Quantitative genetics, inclusive fitness, and group selection. *American Naturalist*, 139(3):540–558.

[Queller, 1994] Queller, D. C. (1994). Genetic relatedness in viscous populations. *Evolutionary Ecology*, 8(1):70–73.

[Queller, 1996] Queller, D. C. (1996). The measurement and meaning of inclusive fitness. *Animal Behaviour*, 51(1):229–232.

[Queller, 2004] Queller, D. C. (2004). Social evolution: kinship is relative. *Nature*, 430(7003): 975–976.

[Queller, 2011] Queller, D. C. (2011). Expanded social fitness and Hamilton's rule for kin, kith, and kind. *Proceedings of the National Academy of Sciences*, 108:10792–10799.

[Queller and Goodnight, 1989] Queller, D. C. and Goodnight, K. F. (1989). Estimating relatedness using genetic markers. *Evolution*, 43:258–275.

[Queller and Strassmann, 2009] Queller, D. C. and Strassmann, J. E. (2009). Beyond society: the evolution of organismality. *Philosophical Transactions of the Royal Society B: Biological Sciences*, 364(1533):3143–3155.

[Quickfall et al., 2015] Quickfall, C. G., Foster, K. R., and Marshall, J. A. R. (2015). Problems with altruism between species. Under submission.

[Raby, 2002] Raby, P. (2002). *Alfred Russel Wallace: A Life*. Princeton University Press.

[Raper, 1984] Raper, K. B. (1984). *The Dictyostelids*. Princeton University Press, Princeton.

[Rapoport, 1967] Rapoport, A. (1967). Exploiter, hero, leader and martyr: the four archetypes of the 2 × 2 game. *Behavioural Science*, 12:81–84.

[Rapoport and Guyer, 1966] Rapoport, A. and Guyer, M. (1966). A taxonomy of 2 x 2 games. *General Systems*, 11:203–214.

[Ratnieks, 1988] Ratnieks, F. L. W. (1988). Reproductive harmony via mutual policing by workers in eusocial Hymenoptera. *American Naturalist*, 132:217–236.

[Ratnieks et al., 2010] Ratnieks, F. L. W., Foster, K. R., and Wenseleers, T. (2010). Darwin's special difficulty: the evolution of "neuter" insects and current theory. *Behavioral Ecology and Sociobiology*.

[Ratnieks and Visscher, 1989] Ratnieks, F. L. W. and Visscher, P. K. (1989). Worker policing in the honeybee. *Nature*, 342(6251):796–797.

[Riley and Wertz, 2002] Riley, M. A. and Wertz, J. E. (2002). Bacteriocins: evolution, ecology, and application. *Annual Reviews in Microbiology*, 56(1):117–137.

[Robertson, 1966] Robertson, A. (1966). A mathematical model of the culling process in dairy cattle. *Animal Production*, 8(95):95–108.

[Robertson, 1968] Robertson, A. (1968). The spectrum of genetic variation. In Lewontin, R. C., editor, *Population Biology and Evolution*, pages 5–16. Syracuse University Press.

[Ross et al., 2013] Ross, L., Gardner, A., Hardy, N., and West, S. A. (2013). Ecology, not the genetics of sex determination, determines who helps in eusocial populations. *Current Biology*, 23: 2383–2387.

[Rousset, 2004] Rousset, F. (2004). *Genetic Structure and Selection in Subdivided Populations*. Princeton University Press.

[Rousset and Lion, 2011] Rousset, F. and Lion, S. (2011). Much ado about nothing: Nowak *et al.*'s charge against inclusive fitness theory. *Journal of Evolutionary Biology*, 24(6):1386–1392.

[Rousset and Roze, 2007] Rousset, F. and Roze, D. (2007). Constraints on the origin and maintenance of genetic kin recognition. *Evolution*, 61(10):2320–2330.

[Russell and Hatchwell, 2001] Russell, A. F. and Hatchwell, B. J. (2001). Experimental evidence for kin-biased helping in a cooperatively breeding vertebrate. *Proceedings of the Royal Society B: Biological Sciences*, 268:2169–2174.

[Samuelson, 1998] Samuelson, L. (1998). *Evolutionary Games and Equilibrium Selection*. MIT Press.

[Samuelson, 1954] Samuelson, P. A. (1954). The pure theory of public expenditure. *Review of Economics and Statistics*, 36(4):387–389.

[Santos et al., 2003] Santos, J. C., Coloma, L. A., and Cannatella, D. C. (2003). Multiple, recurring origins of aposematism and diet specialization in poison frogs. *Proceedings of the National Academy of Sciences*, 100(22):12792–12797.

[Schwartz, 2000] Schwartz, J. (2000). Death of an altruist. *Lingua Franca: The Review of Academic Life*, 10(5):51–61.

[Seeley, 1985] Seeley, T. D. (1985). *Honeybee Ecology: A Study of Adaptation in Social Life*. Princeton University Press.

[Seger, 1981] Seger, J. (1981). Kinship and covariance. *Journal of Theoretical Biology*, 91(1):191–213.

[Segerstrale, 2013] Segerstrale, U. (2013). *Nature's Oracle: The Life and Work of W. D. Hamilton*. Oxford University Press.

[Shipley, 2002] Shipley, B. (2002). *Cause and Correlation in Biology: A User's Guide to Path Analysis, Structural Equations and Causal Inference*. Cambridge University Press.

[Simon, 1956] Simon, H. A. (1956). Rational choice and the structure of the environment. *Psychological Review*, 63(2):129–138.

[Skyrms, 2004] Skyrms, B. (2004). *The Stag Hunt and the Evolution of Social Structure*. Cambridge University Press.

[Smith et al., 2010] Smith, J., van Dyken, J. D., and Zee, C. (2010). A generalization of Hamilton's rule for the evolution of microbial cooperation. *Science*, 328:1700–1703.

[Smukalla et al., 2008] Smukalla, S., Caldara, M., Pochet, N., Beauvais, A., Guadagnini, S., Yan, C., Vinces, M. D., et al. (2008). *FLO1* is a variable green beard gene that drives biofilm-like cooperation in budding yeast. *Cell*, 135(4):726–737.

[Speed, 2001] Speed, M. P. (2001). Can receiver psychology explain the evolution of aposematism? *Animal Behaviour*, 61(1):205–216.

[Spencer, 1864] Spencer, H. (1864). *The Principles of Biology*, volume 1. Williams and Norgate.

[Strassman et al., 2011] Strassman, J. E., Page, R. E., Robinson, G. E., and Seeley, T. D. (2011). Kin selection and eusociality. *Nature*, 471:E5–E6.

[Strassmann et al., 2000] Strassmann, J. E., Zhu, Y., and Queller, D. C. (2000). Altruism and social cheating in the social amoeba *Dictyostelium discoideum*. *Nature*, 408(6815):965–967.

[Sugden, 1986] Sugden, R. (1986). *The Economics of Rights, Co-operation and Welfare*. Blackwell, Oxford.

[Taborsky, 2013] Taborsky, M. (2013). Social evolution: reciprocity there is. *Current Biology*, 23(11):R486–R488.

[Taylor, 1990] Taylor, P. D. (1990). Allele-frequency change in a class-structured population. *American Naturalist*, pages 95–106.

[Taylor, 1992a] Taylor, P. D. (1992a). Altruism in viscous populations—an inclusive fitness model. *Evolutionary Ecology*, 6(4):352–356.

[Taylor, 1992b] Taylor, P. D. (1992b). Inclusive fitness in a homogeneous environment. *Proceedings of the Royal Society B: Biological Sciences*, 249(1326):299–302.

[Taylor et al., 2007a] Taylor, P. D., Day, T., and Wild, G. (2007a). Evolution of cooperation in a finite homogeneous graph. *Nature*, 447(7143):469–472.

[Taylor and Frank, 1996] Taylor, P. D. and Frank, S. A. (1996). How to make a kin selection model. *Journal of Theoretical Biology*, 180(1):27–37.

[Taylor and Jonker, 1978] Taylor, P. D. and Jonker, L. B. (1978). Evolutionary stable strategies and game dynamics. *Mathematical Biosciences*, 40(1):145–156.

[Taylor et al., 2007b] Taylor, P. D., Wild, G., and Gardner, A. (2007b). Direct fitness or inclusive fitness: how shall we model kin selection? *Journal of Evolutionary Biology*, 20(1):301–309.

[Tinbergen, 1963] Tinbergen, N. (1963). On aims and methods of ethology. *Zeitschrift für Tierpsychologie*, 20(4):410–433.

[Traulsen, 2010] Traulsen, A. (2010). Mathematics of kin- and group-selection: formally equivalent? *Evolution*, 64(2):316–323.

[Trivers, 1971] Trivers, R. L. (1971). The evolution of reciprocal altruism. *Quarterly Review of Biology*, 46:35–57.

[Trivers and Hare, 1976] Trivers, R. L. and Hare, H. (1976). Haploidploidy and the evolution of the social insects. *Science*, 191(4224):249–263.

[Tucker, 1950] Tucker, A. W. (1950). A two-person dilemma. In *Readings in Games and Information*, pages 7–8. Blackwell, Oxford.

[van Cleve and Açkay, 2014] van Cleve, J. and Açkay, E. (2014). Pathways to social evolution: reciprocity, relatedness and synergy. *Evolution*, 68:2245–2258.

[van Veelen, 2005] van Veelen, M. (2005). On the use of the Price equation. *Journal of Theoretical Biology*, 237(4):412–426.

[van Veelen, 2009] van Veelen, M. (2009). Group selection, kin selection, altruism and cooperation: when inclusive fitness is right and when it can be wrong. *Journal of Theoretical Biology*, 259: 589–600.

[van Veelen et al., 2012] van Veelen, M., García, J., Sabelis, M. W., and Egas, M. (2012). Group selection and inclusive fitness are not equivalent; the Price equation vs. models and statistics. *Journal of Theoretical Biology*, 299:64–80.

[Von Neumann and Morgenstern, 1944] Von Neumann, J. and Morgenstern, O. (1944). *Theory of Games and Economic Behavior*. Princeton University Press.

[Wade, 1985] Wade, M. J. (1985). Soft selection, hard selection, kin selection, and group selection. *American Naturalist*, 125(1):61–73.

[Weibull, 1995] Weibull, J. W. (1995). *Evolutionary Game Theory*. MIT Press.

[Weismann, 1903] Weismann, A. (1903). *The Evolution Theory*. Edward Arnold, London.

[Weiss, 2006] Weiss, N. A. (2006). *A Course in Probability*. Pearson Addison Wesley.

[Wenseleers and Ratnieks, 2006a] Wenseleers, T. and Ratnieks, F. L. W. (2006a). Comparative analysis of worker reproduction and policing in eusocial Hymenoptera supports relatedness theory. *American Naturalist*, 168:E163–E179.

[Wenseleers and Ratnieks, 2006b] Wenseleers, T. and Ratnieks, F. L. W. (2006b). Enforced altruism in insect societies. *Nature*, 444(7115):50.

[West, 2009] West, S. A. (2009). *Sex Allocation*. Princeton University Press.

[West et al., 2007a] West, S. A., Diggle, S. P., Buckling, A., Gardner, A., and Griffin, A. S. (2007a). The social lives of microbes. *Annual Review of Ecology, Evolution and Systematics*, 38: 53–77.

[West and Gardner, 2010] West, S. A. and Gardner, A. (2010). Altruism, spite, and greenbeards. *Science*, 327(5971):1341–1344.

[West et al., 2007b] West, S. A., Griffin, A. S., and Gardner, A. (2007b). Social semantics: altruism, cooperation, mutualism, strong reciprocity and group selection. *Journal of Evolutionary Biology*, 20(2):415–432.

[West et al., 2001] West, S. A., Murray, M. G., Machado, C. A., Griffin, A. S., and Herre, E. A. (2001). Testing Hamilton's rule with competition between relatives. *Nature*, 409(6819):510–513.

[Williams, 1957] Williams, G. C. (1957). Pleiotropy, natural selection, and the evolution of senescence. *Evolution*, 11(4):398–411.

[Williams, 1966] Williams, G. C. (1966). *Adaptation and Natural Selection*. Princeton University Press.

[Williams, 1992] Williams, G. C. (1992). *Natural Selection: Domains, Levels and Challenges*. Oxford University Press.

[Williams and Williams, 1957] Williams, G. C. and Williams, D. C. (1957). Natural selection of individually harmful social adaptations among sibs with special reference to social insects. *Evolution*, 11:32–39.

[Wilson, 1975a] Wilson, D. S. (1975a). A theory of group selection. *Proceedings of the National Academy of Sciences*, 72(1):143–146.

[Wilson, 1979] Wilson, D. S. (1979). Structured demes and trait-group variation. *American Naturalist*, 113(4):606–610.

[Wilson et al., 1992] Wilson, D. S., Pollock, G. B., and Dugatkin, L. A. (1992). Can altruism evolve in purely viscous populations? *Evolutionary Ecology*, 6(4):331–341.

[Wilson, 1975b] Wilson, E. O. (1975b). *Sociobiology: The New Synthesis*. Belknap Press.

[Wilson, 1980] Wilson, E. O. (1980). Caste and division of labor in leaf-cutter ants (Hymenoptera: Formicidae: *Atta*). *Behavioral Ecology and Sociobiology*, 7(2):157–165.

[Wilson and Hölldobler, 2005] Wilson, E. O. and Hölldobler, B. (2005). Eusociality: origin and consequences. *Proceedings of the National Academy of Sciences of the United States of America*, 102(38):13367–13371.

[Wilson and Nowak, 2014] Wilson, E. O. and Nowak, M. A. (2014). Natural selection drives the evolution of ant life cycles. *Proceedings of the National Academy of Sciences*, 111:12585–12590.

[Wright, 1921a] Wright, S. (1921a). Correlation and causation. *Journal of Agricultural Research*, 20(7):557–585.

[Wright, 1921b] Wright, S. (1921b). Systems of mating. I. The biometric relations between parent and offspring. *Genetics*, 6(2):111–123.

[Wright, 1922] Wright, S. (1922). Coefficients of inbreeding and relationship. *American Naturalist*, 56(645):330–338.

[Wright, 1932] Wright, S. (1932). The roles of mutation, inbreeding, crossbreeding and selection in evolution. In *Proceedings of the Sixth International Congress of Genetics*, volume 1, pages 356–366.

[Wright, 1969] Wright, S. (1969). *Evolution and the Genetics of Populations. Volume 2*. University of Chicago Press, Chicago.

[Wyatt et al., 2013] Wyatt, G. A. K., West, S. A., and Gardner, A. (2013). Can natural selection favour altruism between species? *Journal of Evolutionary Biology*, 26(9):1854–1865.

[Wynne-Edwards, 1962] Wynne-Edwards, V. C. (1962). *Animal Dispersion in Relation to Social Behaviour*. Oliver and Boyd, Edinburgh.

[Yeh and Gardner, 2012] Yeh, A. Y. and Gardner, A. (2012). A general ploidy model for the evolution of helping in viscous populations. *Journal of Theoretical Biology*, 304:297–303.

[Zanette et al., 2012] Zanette, L. R. S., Miller, S. D. L., Faria, C., Almond, E. J., Huggins, T. J., Jordan, W. C., and Bourke, A. F. G. (2012). Reproductive conflict in bumblebees and the evolution of worker policing. *Evolution*, 66:3765–3777.

INDEX

The letters f or t following a page number denote figures or tables.

Ingram Content Group UK Ltd.
Milton Keynes UK
UKHW010627230723
425610UK00002B/76